多样性的中国湿地 DIVERSE WETLANDS OF CHINA

多样性的中国湿地

DIVERSE WETLANDS OF CHINA

陈建伟 著／摄
CHEN JIANWEI

中国林业出版社
CHINA FORESTRY PUBLISHING HOUSE

图书在版编目（CIP）数据

多样性的中国湿地: 一滴水生态摄影丛书: 汉、英/陈建伟著、摄. —北京: 中国林业出版社，2014.5

ISBN 978-7-5038-7585-4

I. ①多… II. ①陈… III. ①沼泽化地－中国－摄影集 IV. ①P942.078-64

中国版本图书馆CIP数据核字(2014)第155356号

责任编辑：田 红
特约编辑：胡京仁
英文翻译：郭瑜富 夏 军 杨瑷铭 蒋英文 陈肖潇
英文审校：郭瑜富
图片编辑：李南岍 黄晓凤 曲笑含
装帧设计：黄华强
制 作：绿野设计工作室

出版：中国林业出版社（100009 北京西城区刘海胡同7 号）
E-mail：wildlife_cfph@163.com
电话：（010）83225764
印刷：北京华联印刷有限公司
版次：2014年9月第1版
印次：2014年9月第1次
开本：787mm×1092mm 1/12
印张：17 $\frac{2}{3}$
定价：258.00 元

湿地大写意之一 “美丽湿地”
金色的大地，蓝色的水，水在陆地上自由地流淌着，呈现出来的是一幅完美的、与地球万物生存休戚相关的“湿地太极图”。

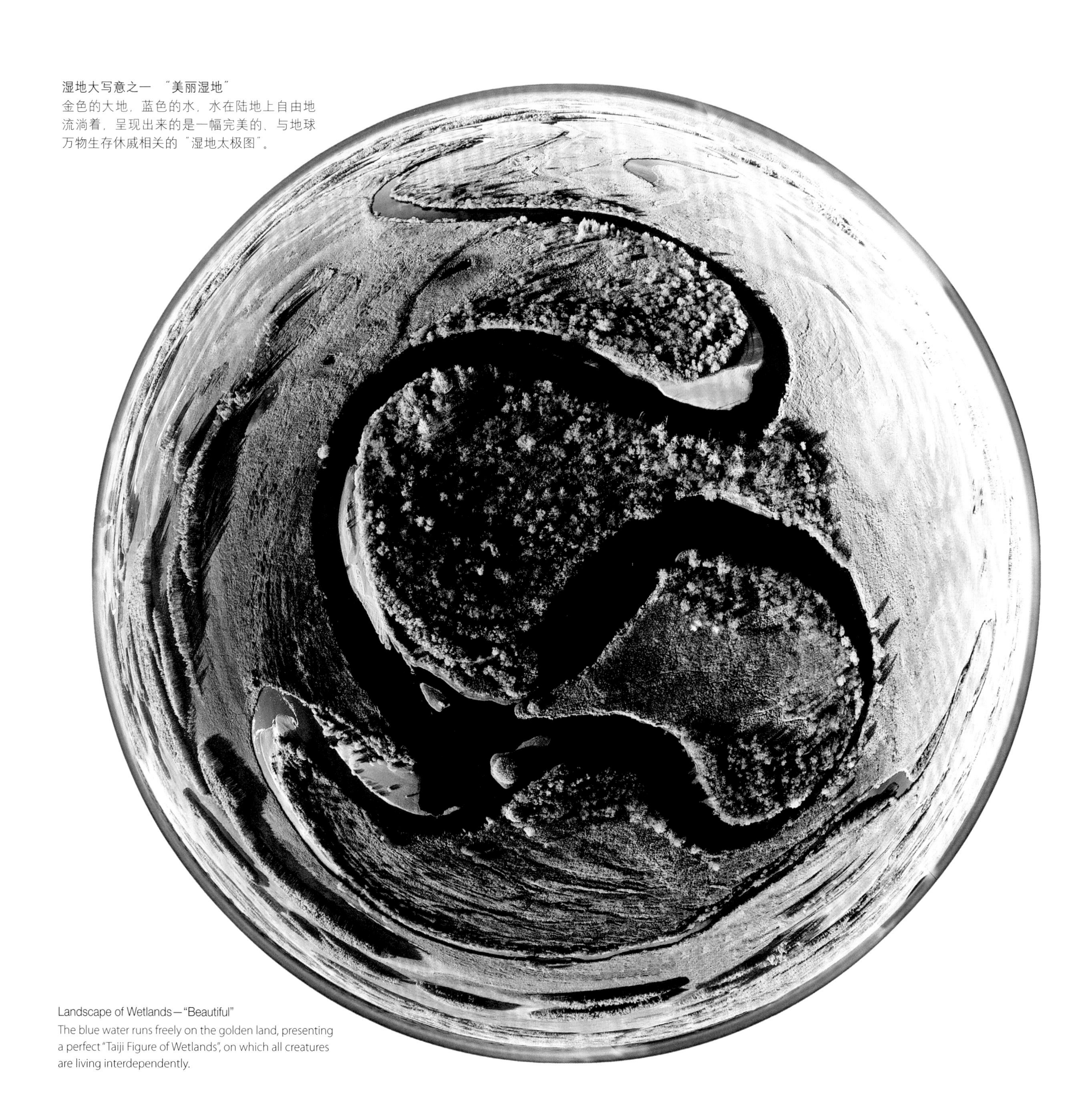

Landscape of Wetlands—“Beautiful”
The blue water runs freely on the golden land, presenting a perfect “Taiji Figure of Wetlands”, on which all creatures are living interdependently.

湿地大写意之二 “混沌湿地”

湿地既不是水体也不是陆地，但湿地既含水体又含陆地，是湿地把水生世界和陆生世界衔接过渡得那么自然和完美。不论是水还是土，也不论是植物还是动物，是低级生物还是高级生物，在这里，都是你中有我、我中有你，众物种相伴相生，形成了一个独特的、统一的湿地生态系统。

Landscape of Wetlands—“Transitional”

Wetland is not water, nor land. It contains both water and land. Wetland is a natural and perfect transition of the aquatic and the terrestrial world. Be it water or earth, plants or animals, lower or higher forms of life, they all live here, interacting with and reinforcing each other. A unique and unified wetland ecosystem is thus formed.

湿地大写意之三 “情意湿地”

湿地千媚百娇、情意绵绵，似走还留、徘徊不定，柔情绰态、回眸频频，始终眷恋着承载她的大地，久久地不愿离去。不论是过去、现在还是将来，蓝色的湿地都是人类文明诞生、发展的摇篮；都是无数地球生命赖以生存、繁衍的天堂。

Landscape of Wetlands—“Affectionate”

Wetland is enchanting and affectionate. It hovers and wanders on land, with tenderness and love. Wetland is attached to the land that bears her, unwilling to part. No matter in the past, at present or in the future, the blue wetlands have always been the cradle for the emergence and development of human civilization, as well as the paradise for the survival and reproduction of the myriad creatures on the earth.

湿地大写意之四 “生命湿地”
有了生命的存在，湿地就有了灵气；有了水鸟的舞蹈，湿地就有了活力；有了动植物的兴衰，湿地就被赋予了特殊的生命意义。湿地是地球上生物种类最丰富的生态系统之一。

Landscape of Wetland—“Alive”
With the existence of creatures, wetland is given vitality; with the dance of waterfowls, wetland is filled with vigor; and with the rise and fall of fauna and flora, wetland is endowed with a special significance of life. Wetland is one of the ecosystems with the richest species of creatures on earth.

序一

湿地是重要的生态系统，也是珍贵的自然资源，不仅具有独特的生态功能，而且具有强大的生产功能，在维持生态平衡、保护生物多样性、促进经济社会可持续发展等方面发挥着重要作用。同时，湿地以其良好的生态环境、优美的自然景观、丰富的生物多样性，吸引着人们休闲旅游、亲近自然，成为普及生态知识、传播生态文化、弘扬生态文明的重要阵地。

我国湿地资源总量位居世界前列，根据第二次全国湿地资源调查结果，全国湿地总面积5360.26万公顷，占国土面积的5.58%。其中，自然湿地面积4667.47万公顷，占全国湿地总面积的87.08%。近年来，国家高度重视湿地保护工作，从政策制度、工程规划、科技支撑、宣传教育、国际合作等方面采取了一系列措施，全国湿地保护工作取得了明显成效。全国已建立577个湿地自然保护区，指定国际重要湿地46处，建立国家湿地公园429个，形成了比较完善的湿地保护体系，为维护生态安全、淡水安全和改善民生福祉发挥了重要作用。但是，由于受到污染、过度捕捞、围垦、外来物种入侵和工程建设等因素的威胁，加之湿地立法滞后、保护投入不足，我国湿地保护还面临着巨大的压力，湿地面积减少、功能退化、生物多样性下降等问题仍然存在。总的看，我国湿地保护与经济社会发展之间的矛盾依然突出，与建设生态文明和美丽中国的要求还不相适应，保护湿地任重道远。

党的十八大明确要求扩大湿地面积，增强生态系统稳定性。这是一项长期而艰巨的任务，需要各级政府和有关部门的高度重视和大力支持，也需要动员全社会力量广泛参与。特别是需要进一步加强宣传教育，引导社会公众科学认识湿地，提高湿地保护意识，自觉做好湿地保护。陈建伟同志多年从事生物多样性保护工作，对湿地保护非常熟悉。结合工作，他致力于科学、真实反映自然环境状况的生态摄影，已经出版了《一滴水生态摄影集》和《多样性的中国森林》两部生态摄影集，得到有关方面和广大读者的好评。陈建伟同志这次出版的《多样性的中国湿地》，以一个生态保护工作者和生态摄影者的视角，用较为丰富的精美图片和简练的文字说明，对我国主要的湿地类型、湿地动植物做了图文并茂的展示，并对加强湿地保护工作进行了深入思考。该摄影集既能给广大读者带来湿地景观美的享受，又有益于提高公众对我国湿地重要性的认识，并引起社会对湿地保护问题的思考。希望能够吸引更多的社会力量积极参与我国林业生态保护和建设事业，争作生态保护的践行者、生态文明的宣传者、生态文化的传播者，为建设美丽中国贡献力量。

国家林业局局长 赵树丛

2014年5月

Forward I

Wetlands, in addition to being critical ecosystems of the planet, are also precious natural resources. They are valuable not only because of their unique ecological functions, but also for their powerful productive functions, playing a significant role in maintaining ecological balance, conserving biodiversity and promoting sustainable economic and social development. In the meantime, with their favorable ambience, appealing natural landscape and rich biodiversities, wetlands are also important recreational places for people to get close to nature, thus making up important platforms for enhancing their eco-awareness and promoting ecological civilization.

China ranks top in the world in its total wetland resources. The Second National Wetland Resources Survey reveals that the total area of wetlands in China amounts to 53.6026 million hectares, accounting for 5.58% of the country's land territory. Among this, 46.6747 million hectares, or 87.08% of the total, are natural wetlands. Over the past years, the Chinese government has attached great importance to wetland conservation, with effective measures taken in policy-making, project planning, technical supports, publicity campaigns and international cooperation to yield impressive achievements. Up to present, 577 wetland nature reserves and 429 state-level wetland parks have been established, together with 46 wetlands that are designated as Ramsar Sites. As a result, a fairly sophisticated wetland conservation system is now in place to safeguard its eco-environment, fresh water supplies, as well as to improve people's livelihood.

However, due to such threats as pollution, overfishing, land reclamation, invasion of alien species and sprawling development programs, complicated by backlagging legislative efforts and insufficient financial inputs, wetland conservation in China is still faced with daunting challenges. Problems like shrinking sizes, functional degradation and diminishing biodiversity still persist. In a word, conflicting demands for wetland conservation and socio-economic development are yet to be tackled. We still have a long way to go to meet the country's strategic goals for promoting ecological civilization and creating a beautiful China.

It is explicitly stated in the documents of the 18th CPC National Congress that wetlands in China should be expanded to enhance its ecosystem stability. This is an arduous task that requires the long-term commitments from governments at all levels, in addition to the extensive, enthusiastic participation of the whole society. Publicity campaigns are particularly needed to public awareness for wetland conservation, and hence mobilizing their interests for such efforts.

Mr. Chen Jianwei has long been actively involved in biodiversity and wetland conservation. Two ecological photo albums of his, *A Single Drop of Water—An Album of Ecological Photography* and *Diverse Forests of China*, have been published and positively commented by both ordinary readers and professionals in relevant institutions. His latest album, highlights the major wetland types and wetland fauna and flora with a rich stock of amazing photos and concise illustrations from the perspectives of a professional conservationist and eco-photographer. Thorough reflections of him on wetland conservation can also be found in the album. It will not only present readers with breath-taking scenes in wetland landscape, but also be conducive to enhanced awareness of the public concerning the significant roles played by wetlands in China. It is hoped that, through this album, more individuals and organizations in the society can be called on to get involved in China's forestry, ecological conservation endeavors, to become the practitioners as well as preachers for eco-conservation and eco-culture, in turn contributing their due shares in building a beautiful China.

ZHAO SHUCONG

Administrator, State Forestry Administration, PRC

May 2014

序二

我和陈建伟先生认识，是在1994年湖南岳阳召开的中国第一次湿地工作会议上，当时来自国家17个部委和国际组织的代表，以及河流（河口）、湖泊、沼泽、海洋、红树林、野生动植物及渔业、水利等方面的专家汇集一堂，共同研讨中国加入《湿地公约》之后如何在“湿地”这个新概念下，保护和合理利用好中国的湿地，作为公约缔约国的中国如何履行好其国际义务和责任。会上陈建伟先生作了《中国湿地的类型及其特点》的报告。

“岳阳会议”至今已经整整二十年了，中国的湿地保护事业从小到大、从弱到强，湿地的保护及合理利用已经得到了各级政府的重视，逐步纳入了国民经济发展计划，资金投入也越来越大，湿地保护为民众所接受并成为了全社会的共识，湿地的科学研究取得了重大进展，各方面工作都取得了很大的成绩。同时，中国的湿地还受到很多威胁和挑战，还存在很多亟待解决的重大问题，还需要我们大家共同努力去加以解决。这个时候，陈建伟先生推出了他的《多样性的中国湿地》生态摄影集，以图文并茂的形式，提纲挈领地将中国多样性的湿地介绍给大家，既有各种典型湿地及物种的美丽展示，又有对湿地的生态思考，集思想性、科学性、艺术性为一体，这是他二十年来始终关注湿地、热爱湿地保护事业的集中体现。

值得一提的是本书有三大亮点。第一个亮点是首次从生态系统的角度来对中国湿地进行区划。原来我们有一个全国湿地八大分区，这个分区更多的是结合行政区划做出的，符合管理的需要。而这本书针对中国自然地理的独特性，综合不同的气候带对湿地生态系统的影响以及中国地势三大阶梯的特点，将全国分为六个大区。从科学性的角度讲，这样的区划更符合不同区域湿地生态系统的特点及规律，更容易进行针对性地分类研究和工作指导，是一种很好的科学尝试。

第二个亮点是立体地表现了湿地生态系统。六个区中，逐次展示该区的河流、湖泊、沼泽、滩涂等湿地类型特点，并加上该区湿地动植物种以及人工湿地、人与湿地的关系等的展示及论述。书的最后部分是生态思考，对于湿地正在受到的威胁和破坏而警醒社会、警醒人们。附件里还有本书涉及的动植物物种名录、保护级别、濒危程度等，便于读者查阅和参考，增强了思想性、知识性和可读性。

第三个亮点是本书特别强调了中国海洋中的湿地，在中国的渤、黄、东、南四大海中分布的数千岛屿上，有着很长的岛屿海岸线，有各种滩涂、岩石海岸及鸟岛等，尤其是提出南海有大量的珊瑚礁、潟湖、环礁等，这里低潮时海平面负6米以上的部分是海洋生物多样性最为富集的地方，是非常美丽又独特的湿地生态系统。这个方面我们以前有所忽视，海洋意识是我们这个民族必须大大提高的。

陈建伟先生的这些努力尝试都很有价值，当然也肯定存在一些考虑不完善、论证不周全的地方，是一家之言也好，是科学争鸣也好，我们都应该给予充分地肯定和鼓励。科学探索需要前人知识的积累和继承，更需要有创新精神，我看，这本书是努力去做了，值得赞赏。

中国科学院院士 刘兴土

2014年5月

Forward II

I made acquaintance with Mr. Chen Jianwei at the first Wetland Conference of China held in Yueyang, Hunan in 1994, which brought together representatives from relevant government departments, international organizations as well as researchers who specialize in studies on rivers, lakes, marsh, ocean, mangrove, wildlife, fishery and water conservancy to discuss how to protect and wisely use wetlands in China under the new concept of "wetlands" after China joins the *Ramsar Convention*. Mr. Chen made a presentation on the Categories and Features of China's Wetlands at the conference.

Over the past two decades since the Yueyang Conference, wetland conservation in China has undergone very impressive progresses, as is demonstrated by the importance attached to the conservation and wise use of wetlands by governments at various levels through incorporating corresponding efforts into the national economic development plan and through the noticeably increased financial inputs. In addition, wetland conservation has become a consensus of the whole society and widely deemed as an essential task by the general public. New breakthroughs have continually been made in scientific researches to bring about tremendous achievements in various aspects. Nevertheless, China is still faced with some serious challenges in wetland conservation that need to be addressed urgently. Published in this context, Mr. Chen Jianwei's ecological album entitled *Diverse Wetlands in China*, with its colorful photographic illustrations and concise explanations, is particularly of significance. The album integrates awesome pictures that highlight the beauty of wetlands and the rich diversity of species inhabiting them with the author's contemplations over China's endeavor in wetlands conservation, thus making it of conceptual, scientific and artistic merits. It is a telling testimony of Mr. Chen's consistent keen concern over wetlands and their conservation over the last two decades.

This album is especially outstanding for the three unique viewpoints contained in it. First, it marks the first attempt in zoning China's wetlands from the perspectives of ecosystems. Previously, wetlands in China were typically divided into eight zones on basis of their geographical location within the country's administrative regions, which catered to the needs of easier administration. This book, in contrast, classifies China into six zones according to the influence of different climatic zones over wetland ecosystems and the "Three-step" topographical structure of China's territory. Judged from scientific viewpoints, such zoning is a very good attempt in that it better represents the features and laws of wetland ecosystems of each particular region and therefore makes it easier to carry out diversified researches that are specially oriented to each region and work out corresponding guidelines for their conservation.

Second, it represents the wetland ecosystem in a three-dimensional perspective. The album not only demonstrates in detail different types of wetlands, such as rivers, lakes, swamps and shoals, of the six wetland zones but also elaborates the relationship between artificial wetlands and human beings and the corresponding flora and fauna species living on it. The last part of the book, "Ecological Reflections", warns the society and the public about the threats and destruction the wetlands are currently faced with. A list of fauna and flora species covered in the book with their grade of protection is also included in the annex for the readers' easy reference, hence making the book both informative and readable.

Third, it underlines the oceanic wetlands of China. On the thousands of islands distributed in Bohai Sea, Yellow Sea, East China Sea and South China Sea, there are lengthy cost lines with various shoals, rocky beaches as well as bird islands. In particular, the book calls our attentions to the numerous coral reefs, lagoons and atolls with the richest marine biodiversity that are located in South China Sea at an altitude of -6 meters in low tides. Not sufficient attention has previously been paid to this type of wetlands. It is necessary for our nation to raise our awareness of ocean

All the endeavors and efforts by Mr. Chen Jianwei in this respect are indeed invaluable, though more thorough considerations and more convincing evidence are needed to better back up some observations made in the Album. Be it his personal preferences or worthy scientific arguments, Mr. Chen's efforts deserve full acknowledgement and due respect from each of us. We need to accumulate and inherit previously available knowledge to carry scientific exploration further on, but innovation is of even higher importance, which, I personally think, is exactly what this Album is aimed at.

LIU XINGTU
Academician of the Chinese Academy of Science
May 2014

前言

《多样性的中国湿地》终于问世了，这是一本倾注了我多年心血的生态摄影集。说起来，我与湿地的渊源要追溯到中国刚刚加入《湿地公约》的二十年前。1994年，中国的第一个湿地会议暨国际会议在岳阳召开，来自国家17个部委，包括湖泊、沼泽、河流、红树林、野生动植物以及水利、农业、海洋、林业等方面的专家汇集在一起，共同研讨中国的湿地如何在公约的框架下结合国情进行保护和建设。这个会上我有幸担任了专家组组长。之后《中国湿地保护行动计划》的编撰、中国的第一次湿地资源调查、中国最大的湿地自然保护区——三江源自然保护区的建立等工作，我都有幸参与其中。二十年来，不论是主管还是不主管湿地保护，我对湿地始终梦牵魂绕，一直不断地支持和关注着她。结合工作，还指导了多名湿地方面的硕士生、博士生，撰写了几十篇有关湿地的论文。

二十年来，中国的湿地保护事业经历了“湿地”新概念下的聚集，起步阶段的艰难，发展过程的曲折，到现在得到社会的普遍认可和民众保护意识的提高，是在克服了诸多困难中前进的。能有机会参与和伴随中国湿地保护事业一起前进，我感到无比的荣幸和自豪，曾经走过的每一块湿地都给了我营养和激情，祖国的大好河川给了我启迪、震撼和思考，这是我要出这本书的原始冲动。

湿地是伟大的，她是人类文明诞生和发展的摇篮；湿地是博爱的，她为地球无数生命提供了优良的栖息环境；湿地是谦逊的，她不争不要，水只往低处流，千回百转奔向大海。滴水成溪，汇溪为河；陡峭成瀑，舒缓为湖；平静踌躇成沼，拥万千生物为怀。我们爱她，就应该关心她、呵护她，当我们看到她受到摧残时，怎能不奋起、不呼喊、不战斗！？

研究和反映中国湿地的重要专著和成果有很多，在这里，我采用了自己较擅长的生态摄影形式，来反映和思考多样性的中国湿地。整理着多年积累的湿地资料和图片，我的体会是：与可以“立起来”的森林相比，湿地是“平躺着”的，在湿地面前，人往往显得渺小和不知所措。湿地是一个相当复杂的生态系统，并以河流、湖泊、沼泽、滩涂等多种形式出现，由于它所处的自然地理区域不同，它的生态内涵也就会不同。我们看湿地不能只局限于“一地一水”，不能“只见水面，不见全面”。怎样从湿地生态系统的角度来认识中国湿地并进行区划，采用什么样的照片来反映等问题都一一摆在面前，而这些问题的解决并无先例可循。

湿地具有多种的生态功能，在与人、与生物、与其他生态系统的关系方面都极为复杂，比起每一块森林、草原、荒漠来讲，每一块湿地都显得更加敏感和脆弱。在湿地生态系统的保护和利用越来越受到人类社会关注的今天，在全球性水资源危机越来越紧迫的形势下，为弘扬生态文明、建设可持续发展的美丽中国，保护和建设好湿地是我们这一代人不可回避的重大历史责任。我对这些问题的认识，将在下面的导言、篇章和生态思考中配合图片进行简述。当然，我个人的认识是粗浅的，缺点错误肯定不少，在“抛砖引玉”的同时，我诚挚地欢迎大家给予批评和指正。

这本生态摄影集的出版，得到了中国生物多样性保护与绿色发展基金会、国家林业局湿地保护管理中心以及宣传中心的大力支持；国家林业局赵树丛局长、中国科学院刘兴土院士为本书分别作序，国家林业局党组成员陈述贤给予了最热情的鼓励和支持，宣传中心程红主任，湿地保护管理中心马广仁主任、严承高、袁继明副主任，鲍达明、方艳等处长给予了大力的支持和宝贵的建议；特别要提到的是，冯宗炜院士、吕宪国、崔丽娟、雷光春、张明祥等专家在业务上给予了我很多指导和帮助；田红、黄华强二位编辑为本书的出版付出了极大的热情和辛劳。另外，在照片拍摄和资料收集过程中，得到了国家及各省（自治区、直辖市）野生动植物、自然保护区、湿地保护管理部门的大力支持，得到了很多自然保护区、湿地公园第一线人员的无私帮助，得到了农业部渔业局肖放、胡学东副局长等的热情关照，还有我的爱人肖江、女儿陈肖潇给予的亲情关怀，我在此一并表示最最诚挚的感谢！没有你们的无私帮助和热情支持，我是无论如何也不可能完成这本书的。

陳建伟

2014年4月

Preface

After years of hard work, my ecological album, *Diverse Wetlands in China,* is finally in press. My emotional attachment to wetlands can be traced back to 1994, when China joined the *Ramsar Convention*. In the same year, China initiated its first international workshop on wetland protection in Yueyang, Hunan Province, focused on measures for protecting and developing China's wetlands that were in line with both the framework of Ramsar Convention and China's national conditions. Experts from 17 ministries and commissions engaged in lake, marsh, river, mangrove, wildlife, water resources, agriculture, ocean and forestry management participated in the workshop. It was a great honor for me to be appointed as the chair of the expert group. Since then, I have had the honor to be actively involved in a series of other initiatives, including formulating *China Wetlands Protection Action Plan*, conducting China's first survey on wetland resources, establishing San Jiangyuan Nature Reserve, the largest wetland nature reserve in the country. In spite of the changes in my administrative positions and that I am no longer in direct charge of wetland protection, my attention and attachment to wetlands never diminished over the past two decades. In addition to supervising some master and doctorate students that major in wetlands protection, I myself have also published dozens of papers in this field.

The notable achievements that China witnessed in wetland conservation over the last two decades have not come by easily. In spite of the numerous difficulties and challenges in the initiating stage as well as tortuous courses of progress, the concept of wetland conservation has so far won extensive acknowledgement in the society, as demonstrated by the public's increasingly enhanced awareness in this aspect. I take great pride in having had the chance to witness and personally participate in the work. Each of the wetlands that I have personally been to, with their spectacular and thought-inducing scenery, still remains vividly in my mind and offer me unceasing inspirations. Hence comes this present album.

Wetlands are precious resources not only in that they provide favorable habitats for human survival and hence have given birth to grandeur civilizations, but also in that they symbolize the most worthy qualities of human being, among which are humanity, modesty and inclusiveness and many others. Being in the form of streams, rivers, lakes, or swamps, wetlands make up the indispensable home for lives on this planet. Given the critical role they play in nurturing lives and civilizations, there is no reason that we human beings do not exert our utmost in caring and conserving the wetlands against devastation.

There are diverse means, for instance the rich monographs and academic papers that have been published, to document the achievements and progresses that China has made in wetland conservation. I would like to contribute my share by resorting to the means in which I have expertise, namely ecological photographing. As I was combing through the pictures that I had taken over the years, it occurred to me that, unlike the vertical forest stands, the vast wetlands that stretch out horizontally are more awe-inspiring and hence more telling about how humble we human beings are in front of nature. Wetlands, with their complicated ecological systems and rich diversities in forms, also remind people that holistic views should be taken so as to understand the nature as an integrated whole. How to recognize and zone China's wetlands from the angle of wetland ecosystem and how to reflect wetlands through photographs accordingly are issues that we do not have previous experience to draw on. In this regard, we still have a long way to go.

With their rich ecological functions and highly intricate relationship with human beings, wildlife communities as well as other ecosystems, wetlands often tend to be more sensitive and fragile than forest, prairie or desert. In the present context when wetland conservation and utilization are gaining increasing attention from the society and are taken as means for addressing pressing global water resource crisis, it is our obligation to pace up our efforts to put wetland resources under effective protection and sustainable uses so as to promote ecological civilization, hence contributing to China's long-term sustainability. I, with the help of my pictures, will try to elaborate on these issues in the following three sections: Preface, Highlights, and Ecological Reflections. Given my knowledge limits, mistakes are bound to exist in this album, for which the readers' comments and constructive feedbacks will be highly appreciated.

The publication of this album wouldn't have been possible but for the keen supports from the China Biodiversity Conservation and Green Development Foundation (CBCGDF), the Office of Wetland Conservation Management (OWCM) and the Publicity Office of the State Forestry Administration (SFA). I am especially grateful to administrator Zhao Shucong of SFA and Mr. Liu Xingtu, academia of Chinese Academy of Science, who prefaced this Album. I am grateful to Mr. Chen Shuxian, SFA vice administrator for his warmest encouragement and support. Valuable advices were provided by Mr. Cheng Hong, Director-General of the Publicity Office, Mr. Ma Guangren, Mr. Yang Chenggao and Mr. Yuan Jiming, heads of the OWCM, and Mr. Bao Daming and Mrs. Fang Yan, division chiefs of the OWCM. In particular, I would like to extend my gratitudes to Mr. Feng Zongwei, academcian of the Chinese Academy of Engineering, Mr. Lyu Xianguo, Ms. Cui Lijuan, Mr. Lei Guangchun and Mr. Zhang Mingxiang and many other specialists for their academic advices. My thanks also go to Ms. Tian Hong and Mr. Huang Huaqiang for their hard work in publication. Besides, during my tour of picture taking, I have been offered invaluable help by corresponding authorities responsible for national and provincial wildlife conservation, nature reserves or wetland conservation, especially Mr. Xiao Fang and Mr. Hu Xuedong, director generals of the Fishery Bureau of the Ministry of Agriculture. Last but not least, my family, my wife Xiao Jiang and my daughter Chen Xiaoxiao, have each given me their dedicated care and love along this journey. Without their help and support, it will not be possible for me to finish this album.

CHEN JIANWEI

April 2014

目录 Contents

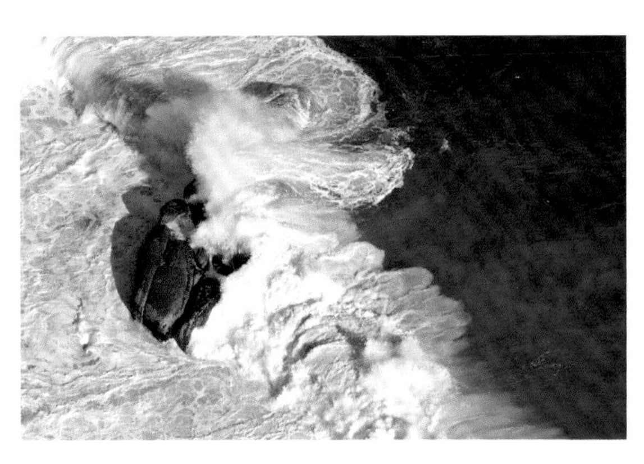

导言

1971年2月2日，在伊朗拉姆萨尔签署的《关于特别是作为水禽栖息地的国际重要湿地公约》（简称《湿地公约》），成为了全球的第一个环境公约。按照《湿地公约》的定义，湿地系指“天然或人工、永久或暂时的沼泽、泥炭地或水域地带，以及蓄有静止或流动的淡水、半咸水或咸水，包括低潮时水深不超过6米的水域。”《湿地公约》将湿地分为三大类——内陆湿地、滨海湿地和人工湿地。由此，我们可以说，湿地是陆地、水体和浅海中各种沼生、湿生区域（包括陆地明水面）的总称，是水陆之间的一个独特的生态系统。

湿地与地球水分循环

水是生命之源，更是湿地之本，没有水，湿地的一切都无从谈起。由于太阳的作用，大气环流把从海洋蒸腾来的水汽，源源不断地输入大陆，加上陆地水分的蒸发，最终变成了云雾、变成了冰川雪山和降水。冰川雪山的融水、大地的降水化成细流涓涓而下，无数小溪小河逐步汇集，一路上“招兵买马”而成为大河。河水从高海拔到低海拔，经高原到平原最终归入大海——于是，一个地球表面水分循环完成了闭合。正是这个伟大的地球水分循环过程作用在大地上，让湿地呈现出河流、湖泊、沼泽、滩涂等各种形态，同时赋予了湿地无穷和强大的生命力。

湿地生态系统

湿地生态系统中的各种生命形式，以生产者、消费者和分解者的角色共同完成湿地的物质循环、能量交换和信息流通。在湿地生态系统里，生产者主要是水中生长的有光合作用的浮游生物，禾草、莎草等挺水植物，浮萍、睡莲等浮水植物以及苦草、金鱼藻、海草等沉水植物组成的各种生物群落，在亚热带热带还有红树、水杉等乔木。消费者主要是在湿地生活的底栖动物、浮游动物、鱼类以及两栖类、爬行类、鸟类、兽类等；而分解者主要是一部分无脊椎动物和微生物等。不同物种间的相互依存、相互制约和有机统一，在复杂食物链中各司其职，维护着生态系统的平衡，共同构成了一个健康和谐的湿地

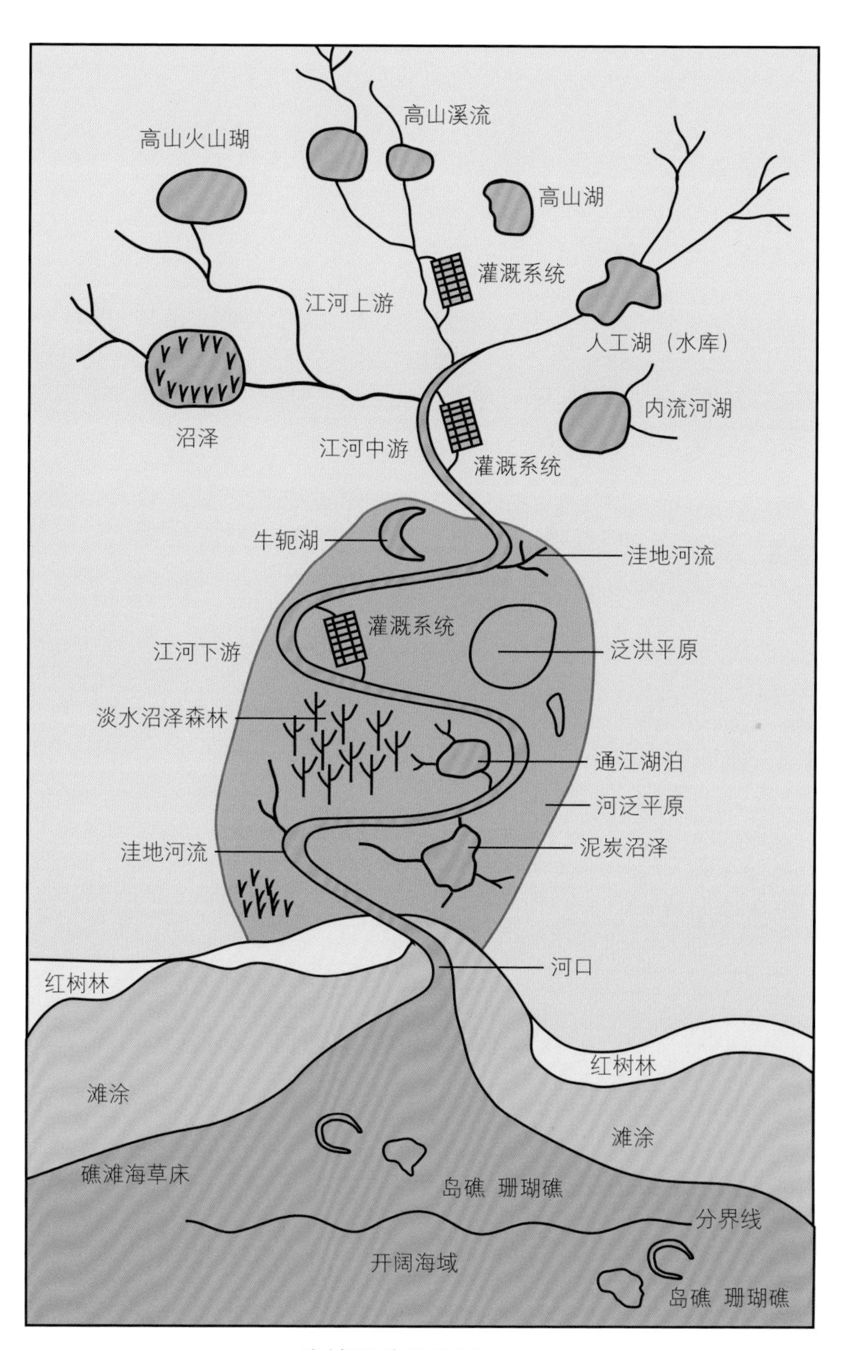

流域湿地示意图

生态系统。因此，我们“看”湿地不仅要从形态、类型、水文、生物、“特别是作为水禽栖息地”的角度来看，更重要的是要从湿地是生态系统的角度来看。我们应该充分认识到：地球的水分循环是一个完整的系统；一个流域的各种各类湿地是相互支撑和相互依存的；流域和流域之间在大气水分循环作用下是相互影响的；依赖湿地生存的各种生命形式是相伴相生的，水禽、鱼类的迁徙更是一个完整的循环链。因此我们“看”湿地，不仅要整体地看、流域性地看，更要从生态系统的角度地来看，不能“只见水面，不看全面”“只知有水就有湿地，不知没有生命就没有湿地（生态系统）”。

多样性的中国湿地

中国具有完整的地球水分循环的各个环节，中国的大地承接着以世界最高峰珠穆朗玛峰领衔的，从喜马拉雅山、昆仑山、唐古拉山等大山下来的水流。作为“中华水塔”、“亚洲水塔”的青藏高原发源着黄河、长江这样的世界级大河和亚洲的大部分河流。在多种不同的气候条件下，在地面径流、天空降水和光、热、土等综合环境因素的共同作用下，水（河）流在地势的低凹处或山口等集水区上形成了各种湖泊；在河湖海边泥沙的不断淤积下形成了滩涂；在平缓洼地上与植物共同作用，以及泥炭的逐渐积累形成了不一样的沼泽；在入海河流和海岸潮汐的共同作用下形成了滨海湿地。这些湿地形成于不同气候、不同纬度、不同海拔、不同光热土组合的环境中，使得依赖其生存的各种不同的动植物种得以生长繁荣，呈现出了一片勃勃生机，最终构成了中国多样性的湿地生态系统。中国湿地生态系统的多样性和完整性是世界其他国家无法比拟的。

中国湿地区划的基础

中国幅员辽阔，经纬度跨越很大，同时地形复杂、地貌多样，自然地理因素极为复杂。要深入理解中国湿地、理解中国湿地的多样性，对中国湿地进行合理的区划是最为重要的基础。合理的区划首先要考虑的就是中国特殊的综合自然地理要素。

研究认为，不论湿地的表现形式是河流、湖泊、沼泽还是滩涂，由于湿地所处气候区域冷暖、干湿的不同组合，它们的内在形式是有很大差异的，而光、热、水、土条件是关键因子。譬如：湿地植物的生长量是由寒温带向热带逐渐增大的，植物残体的分解能力也是由寒温带向热带逐渐增加的。在生长量和分解能力相互作用的条件下，在我国温带的干旱、半干旱区由于生长量小，相对的分解能力强，因此很难形成沼泽。而温带的湿润地区由于温湿的环境条件，生长量大，相对的分解能力弱，就有大量的沼泽分布。在热带，尽管生长量大，但相对分解能力也很强，导致泥炭积累弱，就极少形成沼泽。总的说来，凡是干燥的条件下，无论是寒冷还是温热环境，都抑制沼泽的形成和发展；湿润的条件下，寒湿、冷湿、温湿的水热组合条件都有利于沼泽的形成和发育，只有热湿除外。又如：湖泊的形成和演变由于所在地气候区的不同也表现不同，干燥条件下由于降水稀少，地表径流补给不丰，蒸发强度较大，多为半咸水、咸水湖。而冷湿、温湿地区由于湖泊补水较丰，河湖关系密切，多为淡水湖泊。而影响河流的不同形成、流向的最重要影响因素就是融水、降水、海拔、地形地势等等。

以此认识，无论湿地是以河流形式还是以湖泊、沼泽等形式出现，依赖其生存的微生物、植物、动物以及沼泽土或泥炭土的形成，是无时无刻不受到其所在地气候条件的长期影响和制约的。有“湿地精灵”之称的水鸟是与各地的气候、水文、植被等自然地理特点相适应的。如：温带水鸟的种类以夏候鸟和旅鸟占优势；亚热带和热带水鸟的种类则以冬候鸟和留鸟占优势。很多水鸟是在温带繁殖，到亚热带、热带越冬的。

综上所述，气候环境因素对生态系统的塑造起着关键性的作用，是我们区划湿地生态系统必须考虑的前提。

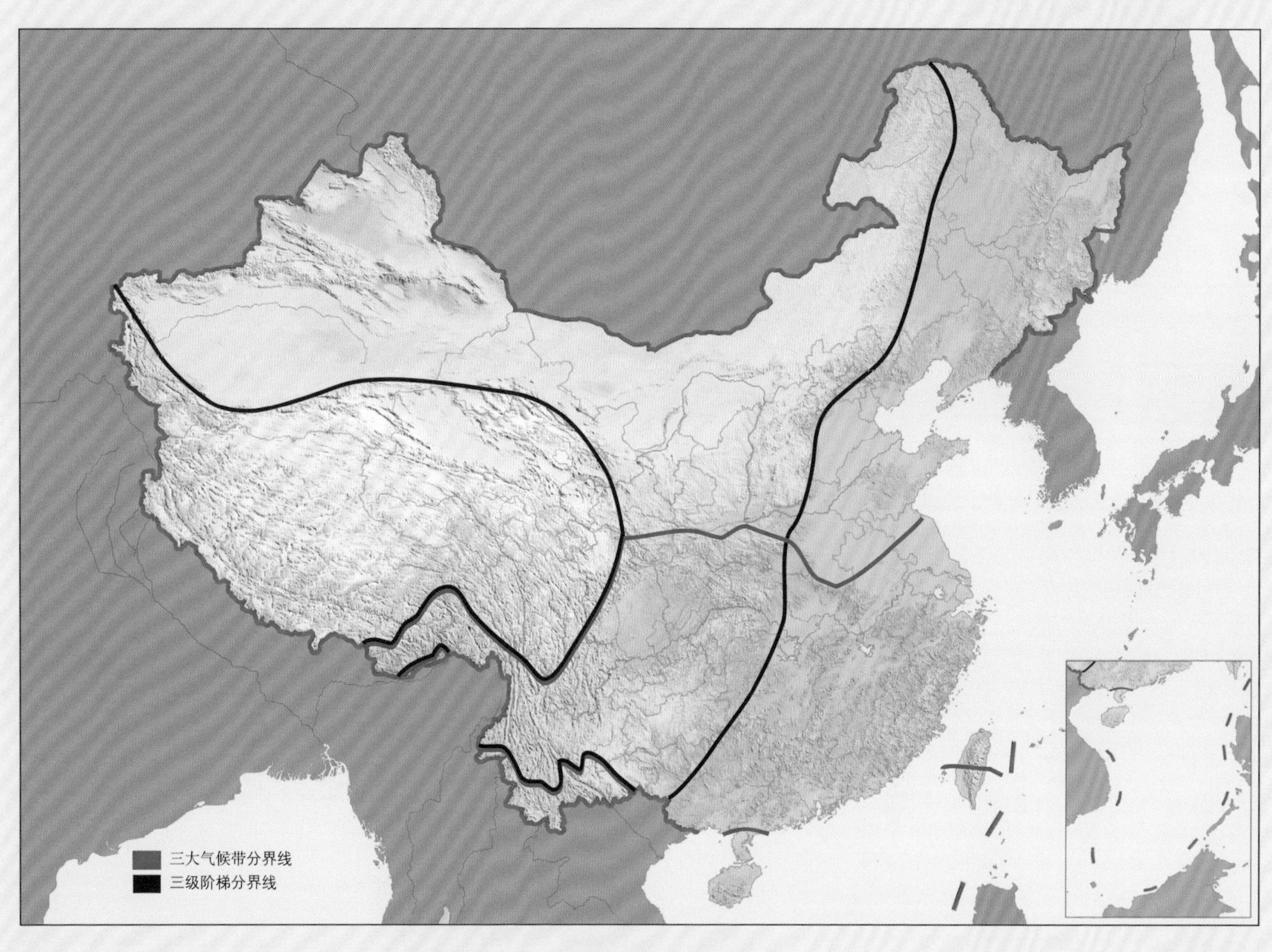

中国湿地区划图

● 中国温带与亚热带的划分，从东部的淮河秦岭一线，再沿青藏高原东缘向南，进入云南境后往西，沿着雅鲁藏布江大拐弯北向上凸起后再回来，然后继续往西延伸，在不丹东北角出境。这条线与世界动物地理区划中的大界——古北界与东洋界在中国的走向基本一致，按此方案，青藏高原及以北地区除了喜马拉雅东段南坡属亚热带、最南端为热带外，其余都属于温带。

● 中国亚热带与热带的划分，从台湾的北回归线往西偏南方向，到广东的雷州半岛北，到云南南部一线出境缅甸，再从藏东南的南部划出。

● 中国第一级阶梯和第二级阶梯的分界线，从昆仑山西段往东，经阿尔金山、祁连山，再沿青藏高原东缘向南，进入云南境后往西，沿着雅鲁藏布江大拐弯北向上凸起后再回来，然后继续往西延伸，在不丹东北角出境。

● 中国第二级阶梯和第三级阶梯的分界线，从大兴安岭往南再偏西方向，走太行山、巫山、雪峰山再沿云贵高原东、南边缘及藏东南的南部出境。

中国湿地区划

中国气候区划的特殊性在于不仅仅要考虑地带性，还要考虑海拔高度的不同所带来的变化。迄今为止，我国还没有进行过基于综合自然地理的湿地区划，为了更加强调是从生态系统而不仅仅是从湿地类型或行政管理的角度上来认识湿地，本书提出的中国湿地区划，紧紧抓住的是两个要素：一是气候带，二是海拔差。

中国国土在南北尺度上，跨越了温带和热带，由于中国的亚热带非常特殊，且面积占到了陆域国土面积的1/4。因此，我们将亚热带单独提出，而按温带、亚热带、热带三个气候带来进行论述。在东西尺度上，由于海拔的巨大差异，产生了地势的不同阶梯。而中国地势的不同阶梯大大影响了气候带在纬度上的“平行分布”，因而我们就有了以水平地带性为依据，在此基础上再表述垂直带谱特征的基带地带性原则。

中国历来各种不同的区划中，在地势较为平坦的东部，温带、亚热带、热带三个区的划法基本是一致的，争议的关键是在西部地区。由于中国的西部地形地貌复杂、相对高差很大，尤其是地势阶梯的存在，西部地区各气候带的分界线的确很难得到公认。经过多方案比较研究，本书以中国气候区划为蓝本，以黄秉维先生的综合地理区划方案为基础，结合吴征镒先生关于中国植被类型的划分，参考中国动物地理区划，形成了现在的中国温带、亚热带和热带的划分，以此作为本书湿地区划的第一个重要依据。

研究证明，中国湿地生态系统在不同地势阶梯环境中的形成以及相关湿地物种方面均有较大差异性，作为湿地基本表现形式的河流、湖泊、沼泽、滨海湿地等在不同地势阶梯上也有各自特点。本书依据孙鸿烈先生的《中国生态系统》中关于中国地势水平分布的三大阶梯划分方法，以青藏高原区为第一级阶梯，蒙新高原区和云贵高原区为第二级阶梯，东北、华北、长江中下游平原以及东南丘陵区为第三级阶梯，以此作为本书湿地区划的第二个依据。

这样，在中国湿地地图上，三个气候带的两条分界线以及三级阶梯的两条分界线就将中国湿地区划为六个部分，分别是：

一，青藏高原湿地区

二，蒙新高原湿地区

三，云贵高原湿地区

四，东北、华北平原湿地区

五，华东、华中、华南平原丘陵湿地区

六，西南、华南山地湿地区

当然，在此基础上还可以再细分亚区，比如东北、华北平原湿地区就可以分为东北亚区和华北亚区，东北亚区还可以按寒温带、中温带、暖温带再做第三级区划等等，但本书仅就一级区划来展开。各个湿地分区的不同特点及做出这样区划更进一步的理由，本书将在各章中继续深入阐述。

人类共同的湿地

湿地不仅为人类提供大量食物、原料和水资源，而且在维持生态平衡、补充地下水、控制土壤侵蚀、保持生物多样性、涵养水源、蓄洪防旱、护岸保堤、降解污染、固碳释氧、调节气候等方面均起到非常重要的作用。湿地以占地球近9%的面积，存储着陆地生物圈35%的碳，养育着地球40%的物种，提供了人类 96%以上的可利用淡水。作为陆地与水体之间的过渡区和生态交错区，湿地生态系统比起其他生态系统来讲更具有它的特殊性。由于水体的流动和水分的循环，分散的一地一处的湿地被高度地联系在了一起。因此，认识湿地、保护湿地不能只局限于某地某区，需要的是系统性和全局性的理念和统筹，需要不同区域人们的共同参与和合作，需要全人类的共同努力。这也就是为什么《湿地公约》能成为全球首个环境公约，得到国际间共识的原因。

人类的文明历程始终伴随着对湿地的依赖、利用和保护。比起其他生态系统来说，湿地生态系统的利用和保护面临着更多的利益平衡和更复杂的问题。但无论如何，我们都应该拿出更大的智慧和决心来，因为每一块湿地都不仅仅只是属于它所在的某个地区或某个人群的，湿地生态系统的保护是没有地界、国界的，保护湿地是我们人类共同的责任。

Introduction

On Feb 2, 1971, the *Convention on Wetlands of International Importance, especially as Waterfowl Habitat* (the "*Ramsar Convention*"), adopted in Ramsar, Iran, became the first world convention on environments. Under the *Ramsar Convention*, wetlands are defined as "areas of marsh, fen, peatland or water, whether natural or artificial, permanent or temporary, with water that is static of flowing, fresh, brackish or salt, including areas of marine water the depth of which at low tide does not exceed six meters." The *Ramsar Convention* categorizes wetlands into three kinds—inland, coastal and artificial. Therefore, we can say that the wetland is a general name of various marsh and wet regions (including water surface on land) on land, water and shallow seas. It is a unique ecosystem between water and land.

Wetlands and Earth's Water Circulation

Water is the source of life, and the basis of wetlands. Without water, there will be no wetlands. Thanks to the sun, the atmospheric circulation transports the vapor rising from the ocean to the land. Coupled with the evaporation of water on land, clouds, mist, glaciers, and snow on mountains are formed. The water melts from glaciers and snow on mountains; and the rain falls on land. Drop by drop, water becomes streams, and streams merge into large rivers. Rivers flow from plateaus down to plains, and finally, into the ocean. In this way, the water circulation on the surface of the earth is completed. It is this great water circulation on earth that gives wetlands various forms, rivers, lakes, marshes, and mud flats, and bestows wetlands with limitless and powerful vitality.

Wetland Ecosystem

In the wetland ecosystem, various forms of life, as producers, consumers and decomposers, complete the wetland's material circulation, energy exchange and information flow together. Producers are various biocenosis with photosynthesis; they are made up of plankton, emergent aquatic plants (such as grasses and *Cyperus rotundus*), floating plants (such as *Lemna minor* and *Nymphaea tetragona*) and submerged plants (such as *Vallisneria natans*, *Ceratophyllum demersum*, and seagrasses). In subtropics, producers also include arbor trees, for example, mangroves and *Metasequoia glyptostroboides*. Consumers are benthonic animals, zooplankton, fishes, amphibians, reptiles, birds and beasts. Decomposers including some invertebrates, microorganisms and so on. Different life species are interdependent. They restrain each other, and form an organic unity. In the complicated food chain, they play their respective roles, maintain the balance of the ecosystem, and construct a healthy and harmonious wetland ecosystem. Hence, we should not only focus on

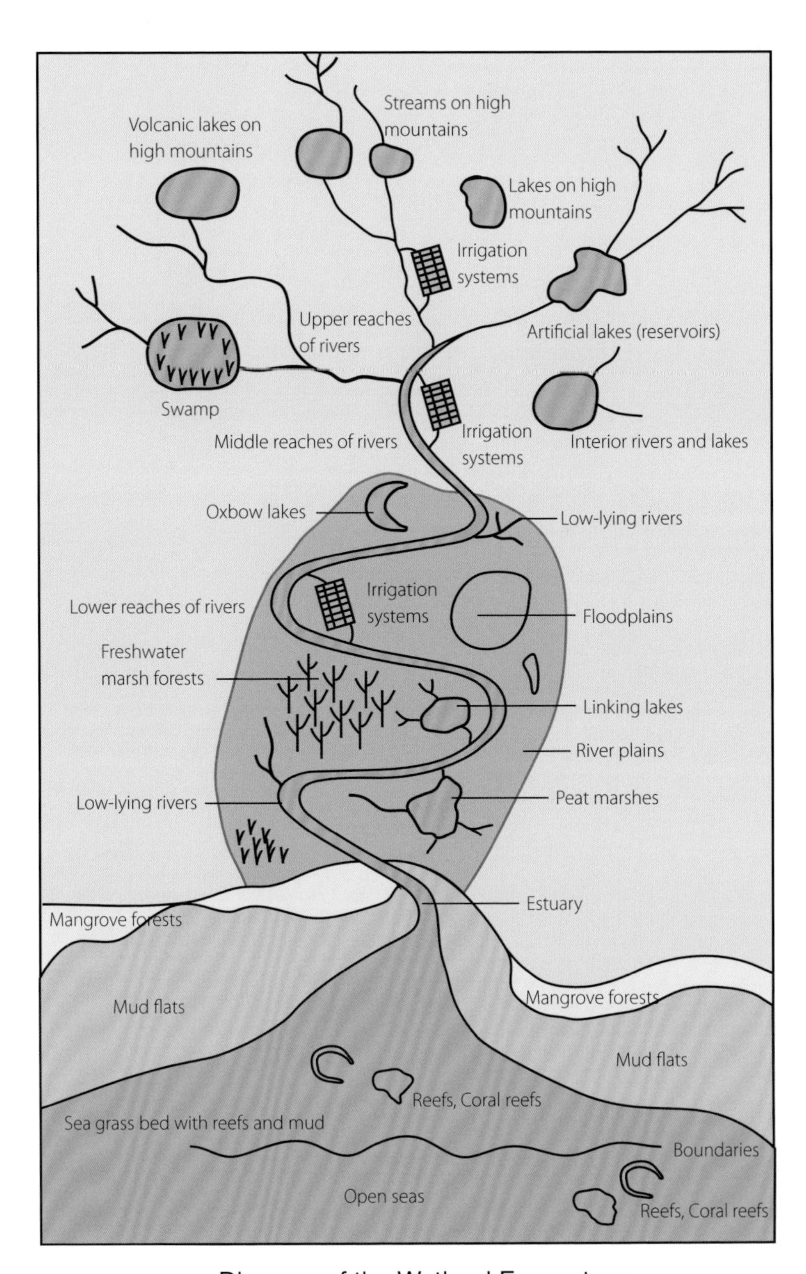

Diagram of the Wetland Ecosystem

wetlands' forms and types, hydrology and organisms, or the "waterfowl habitat", we should also bear in mind that the wetland is an ecosystem. We should fully recognize that the earth's water circulation is a complete system; that all kinds of wetlands in a drainage basin are reinforcing and interdependent; that different drainage basins can affect each other under the water circulation in the atmosphere; and that the waterfowl's migration, dependent on wetlands, is a complete circulation chain. Therefore, we should look at the wetland from the holistic perspective of the drainage areas and the ecosystem. We should not "focus only on the surface, but not on the whole". We cannot "only know that where there is water, there are wetlands; but not know that without life, there will be no wetlands (ecosystems)".

Diverse Wetlands of China

China is gifted with a complete water circulation on earth; and on China's land, flows water from the Himalaya Mountains, the Kunlun Mountains, the Tanggula Mountains and other mountains. Known as "China's Water Tower" as well as "Asian Water Tower", the Qinghai-Tibet Plateau is the source of many Asian and the world-famous rivers, such as the Yellow River and the Yangtze River, just to name a few. Under the integrated environmental conditions of runoffs on land, rainfall, light, heat and earth, lakes were formed on low lands, mountain passes and other catchment areas; mud flats through sedimentation by rivers, lakes and oceans; marshes on gentle low-lying lands after interaction with plants and gradual accumulation of peat; and coastal wetlands under the combined action of rivers running into the sea and tides. These wetlands are formed on different latitudes and altitudes, and environments with different light, heat and soil. As a result, various kinds of animals and plants depending on the wetlands thrive and are full of vigor and vitality, constructing China's diverse wetland ecosystems. The diversity and completeness of China's wetland ecosystems is incomparable all over the world.

Basis for the Zoning of China's Wetlands

With a vast territory, China spans a lot in longitudes and latitudes. Meanwhile, China has a complex terrain and diverse landscapes. Thus, it has very complicated natural and geological factors. In order to understand China's wetlands and their diversity, a sound zoning is the most important basis. The first and foremost to be considered is China's special integrated natural and geological factors.

Studies show that as regions where wetlands (rivers, lakes, marshes or coasts) reside have different combinations of temperature and humidity, wetlands can differ greatly. Light, heat, water and soil are critical factors. For example, the growth of plants on wetlands in tropics is faster than that in cold temperate zones; and the decomposition of plant residues in tropics is stronger than that in cold temperate zones. In China's temperate arid and semi-arid zones, the growth is slow, the decomposition is strong, so marshes can hardly be formed. In the temperate humid zones, the growth is fast, the decomposition is weak, and marshes are widely distributed. In tropics, despite the fast growth, the decomposition is strong and the peat's accumulation is weak so almost no marshes can be found. Generally speaking, dry conditions, cold or warm, restrain marshes from development; while humid conditions, frigid, cold and warm, are good for the development of marshes. To take another example, depending on different climates, the development and evolvement of lakes can also differ greatly. In dry conditions, with little precipitation and surface runoffs, the evaporation is intense, and brackish-water lakes and salt-water lakes are formed. In cold or temperate humid zones, with abundant water in lakes, the rivers and lakes are closely connected, and fresh-water lakes can be found. The most important influencing factors of rivers' development and directions are melt water, precipitation, elevation and terrain, etc.

From this perspective, no matter wetlands are in the form of rivers, lakes or marshes, the microorganisms, plants, animals, and the development of boggy soil and peat soil, depending on wetlands, are always under the influence of the climatic conditions of wetlands. Waterfowls, "the spirit of wetlands", are closely related with the local natural and geological characteristics, climate, hydrology, vegetation and so on. For instance, in temperate zones, summer birds and passing migrant birds are dominant; while in sub-tropics, winter birds and resident birds are. Many waterfowls breed in temperate zones, and spend the winter in subtropics and tropics.

In conclusion, climatic and environmental factors play a critical role in shaping the ecosystems, and are the preconditions that must be considered in wetland ecosystems zoning.

China's Wetlands Zoning

In China's climate zoning, zonality is not the only thing to be considered, different altitudes can also bring different changes. Up till now, China has not carried out wetlands zoning based on natural and geological conditions. In order to understand wetlands from the perspective of ecosystems, not just wetland types or administration, the wetlands zoning proposed in this book focus on two elements: climate zones and altitudes.

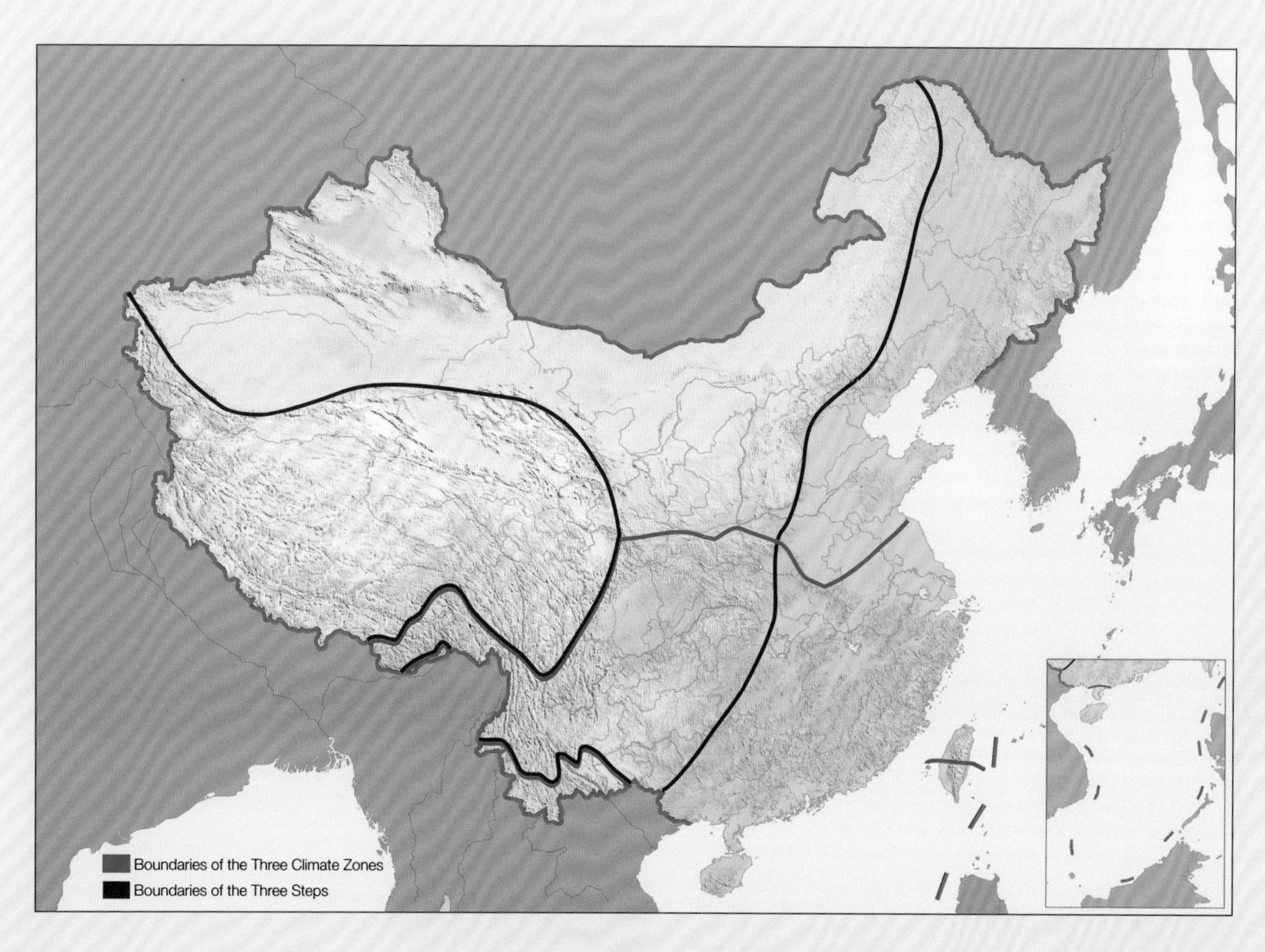

Regionalization Map of China's Wetlands

- The boundary between China's temperate zone and the subtropics starts from the Huai River-Qin Mountains line in the east, stretches south along the eastern edge of the Qinghai-Tibet Plateau, travels west in Yunnan Province, moves north along the big turn of the Yarlung Tsangpo River and returns, extends further west, and leaves China at the northeast corner of Bhutan. This line is basically the same as the world ecozone boundary, between the Palearctic ecozone and the Indomalaya ecozone in China. Accordingly, the area north to the Qinghai-Tibet Plateau is the temperate zone, except the southern slope of the east Himalaya Mountains, which is the subtropics, and the southernmost, which is the tropic.
- The boundary between China's subtropics and tropics stretches southwest from the Tropic of Cancer in Taiwan, extends to the northern Leizhou Peninsula of Guangdong Province, travels in the south Yunnan Province, leaves for Burma, and ends at the southern southeast corner of Tibet.
- The boundary between China's first and second step extends east from the west Kunlun Mountains, travels along the Altun Mountains and the Qilian Mountains, stretches south along the eastern edge of the Qinghai-Tibet Plateau, travels west in Yunnan Province, moves north along the big turn of the Yarlung Tsangpo River and returns, extends further west, and leaves China at the northeast corner of Bhutan.
- The boundary between China's second and third step stretches southwest from the Greater Khingan Range, extends along the Taihang Mountains, the Wu Mountain and the Xuefeng Mountain, and leaves China at the south and east edge of the Yunnan-Guizhou Plateau and ends at the southern southeast corner of Tibet.

In latitudes, China's territory spans the temperate zone and the tropics. As China's subtropics are special and cover 1/4 of China's land area, we consider subtropics as an independent climate zone and discuss it together with the other two climate zones—the temperate zone and the tropic zone. In longitudes, due to the large difference in elevations, China's terrain has different steps. Consequently, it influences the "horizontal distribution" of climate zones in latitudes. Thus, based on the horizontal zonality, we have the base band zonality principle with vertical zonation as characteristics.

In history, China's zoning varies greatly. In the east, which is relatively flat, the zoning of the temperate zone, the subtropics and the tropics is basically the same. The conflict lies in the west. As China's west has a complicated terrain and a relatively large altitude difference, especially the existence of terrain steps, the boundary of climate zones in the west can hardly be agreed upon. After comparative studies, this book, with China's climatic zoning as the blueprint, based on Mr. Huang Bingwei's integrated geographic zoning, combined with Mr. Wu Zhengyi's division on China's vegetation type, and with reference to China's animal geographic zoning, develops the current zoning of China's temperate zone, the subtropics and the tropics. This zoning is the first basis of the wetlands zoning in this book.

Studies show that China's wetland ecosystem differs largely in the development of different terrain environment and the related species on wetlands. As the basic forms of wetlands, rivers, lakes, marshes and coasts have their respective characteristics in different terrain steps. According to Mr. Sun Honglie's Three-step zoning on China's horizontal terrain distribution in *China's Ecosystems*, the Qinghai-Tibet Plateau is the first step, the Inner Mongolia-Xinjiang Plateau and the Yunnan-Guizhou Plateau are the second, the Northeast China, the North China, the middle and lower reaches of the Yangtze River and the Southeast China's hilly areas are the third. This is the second basis of this book's wetland zoning.

On China's wetland map, the two boundaries of the three climate zones and the two boundaries of the three steps divide China's wetlands into six parts, namely:

1. Qinghai-Tibet Plateau Wetland Zone;
2. Inner Mongolia-Xinjiang Plateau Wetland Zone;
3. Yunnan-Guizhou Plateau Wetland Zone;
4. Northeast and North China Plains Wetland Zone;
5. East, Central and South China Plains and Hilly Areas Wetland Zone;
6. Southwest and South China Mountainous Areas Wetland Zone.

Of course, on this basis, sub-zone can also be further divided, for instance, the Northeast China and the North China Plains Wetland zone can be divided into the Northeast China Sub-zone and the North China Sub-zone; and the Northeast China Sub-zone can be divided further according to the frigid-temperate zone, moderate-temperate zone and warm temperate zone. However, this book only discusses the first grade of the zoning. The different characteristics of the wetland zoning and its further reasons will be discussed in depth in the following chapters.

Wetlands as Common Treasures of Mankind

Wetlands not only provide food, raw materials and water for humans, but also play an important role in ecological balance maintenance, groundwater replenishing, soil erosion control, biodiversity maintenance, water conservation, flood storage and drought control, banks and dikes protection, degradation and pollution control, carbon sequestration and oxygen release, and climate regulation. Wetlands, covering 9% of the earth's area, store 35% of carbon, nurture 40% of the species on earth, and provide over 90% of water for humans. As the transition between land and water, the wetland ecosystem is different from other ecosystems. Due to the flow and circulation of water, wetlands in different places are closely related together. Therefore, in order to understand and protect wetlands, we should not just focus on some places. It is important to have a systematic and holistic view. People from different parts of the world need to participate, cooperate and work together. That is why the *Ramsar Convention* is the first world convention on environments, and is agreed by the international society.

The historic progress of the mankind civilization is accompanied by people's reliance on, and their development and exploitation of wetlands. Compared with other ecosystems, the wetland ecosystem is faced with more conflicts of interest and more complicated problems in its protection. Under whatever circumstances, we should show our wisdom and resolution. No wetland only belongs to one place or one group of people. The protection of wetland ecosystems has no boundaries. It is the common responsibility of mankind to protect wetlands.

1 青藏高原湿地区

Qinghai-Tibet Plateau Wetland Zone

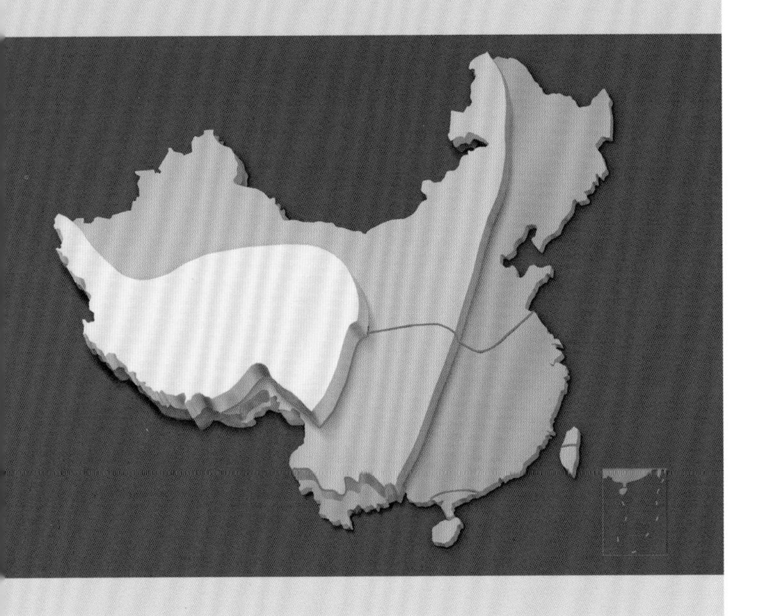

(前页) 雪岭冰川　沮泽沼滂 · 西藏喜马拉雅山东段
来自海洋源源不断的暖湿气流，在高寒的青藏高原生成了无数的冰川雪山，冰川雪山融水化成细流涓涓而下，不断滋润着贫瘠的冲积扇，形成了小溪、河流和湿草甸，生命的湿地自此铺了开来。

(Previous Page) Snow Ranges, Glaciers and Marshes — East Himalaya Mountains, Tibet
Warm and humid currents, rising from the ocean, become glaciers and snow mountains on the cold Qinghai-Tibet Plateau. The ice and snow melt into streams and flow down, nurturing the barren alluvial fan. Streams, rivers and marshy meadows are developed, and life starts.

青藏高原湿地区

这里是中国地势的第一级阶梯，青藏高原湿地区占到陆域国土面积的1/4，平均海拔4000米以上，有“世界屋脊”之称。地球上中低纬度地区的冰川主要集中在这里，是我国以及亚洲许多大河的发源地。长江和黄河发源于此并贯穿中国一、二、三级阶梯，国际河流澜沧江（下游为湄公河）、怒江（下游称萨尔温江）、狮泉河（下游称印度河）、雅鲁藏布江（下游称布拉马普特拉河）等均发源于此。流往我国蒙新高原的塔里木河、黑河、石羊河等也在此区发源。这里气候区属温带，气候严寒而干旱，虽降水稀少，但冰雪融水是河流、湖泊的主要补给形式，水力资源相当丰富，有着地球上海拔最高、数量最多、面积最大的湖群。由于低纬度太阳强烈的蒸发作用，湖水入不敷出，大都发育为内陆湖，只有少数为外流湖，因此多以盐湖和咸水湖为主。草甸草原地带的沼泽主要位于高原的江河源区或者近河源区，发育在河漫滩、洼地，而泥炭沼泽土、草甸沼泽土为湿地主要土壤类型，最具特点的沼泽类型主要为西藏嵩草沼泽和木里苔草沼泽。在其东北部的若尔盖，寒冷湿润的地面环境条件形成了中国面积最大的沼泽泥炭地。湿地里浮游生物、底栖动物种类少、数量低，大多分布一些耐盐碱的鱼类。青海湖、扎陵湖、鄂陵湖、纳木错、若尔盖、雅鲁藏布江中游等都有大量的水鸟栖息，数量甚多但种类较少，主要有斑头雁、渔鸥、棕头鸥、普通鸬鹚等。若尔盖是全球唯一的高原珍稀鹤类——黑颈鹤的主要繁殖地，雅鲁藏布江流域中段是其主要越冬地之一。

在这个区域里，人烟稀少，人为利用湿地不多，绝大多数湿地都是以自然状态存在的。但该区域的湿地状况与全国、与亚洲关系却是异常紧密的，影响是重大深远的。这里发源的数条著名大河源远流长，影响着亚洲近20亿人口的生活，这里的湿地水文、气候的微小变动都会波及全国以至亚洲大地，是我们要竭尽全力来保护的“中华水塔”、“亚洲水塔”。

(从左至右)
珠峰下的杂嘎曲 · 西藏定日
水天之间 · 雅鲁藏布江中游湖泊
鲜花的沼泽 · 青海玉树
普通鸬鹚 · 青海湖鸟岛
圣湖神鱼 · 四川巴塘措普湖

(From Left to Right)
Zagaqu under the Mount Everest — Tingri, Tibet
Between water and heaven — Lake on the middle reaches of the Yarlung Tsangpo River
Marsh with Flowers — Yushu, Qinghai
Great Cormorants — Bird Island, Qinghai Lake
Fish in the Sacred Lake — Cuopu Lake, Batang, Sichuan

Qinghai-Tibet Plateau Wetland Zone

As the first step of China's terrain, the Qinghai-Tibet Plateau covers 1/4 of China's land area. With the average elevation of over 4,000 meters, the Qinghai-Tibet Plateau is known as "the roof of the world". Most glaciers at the middle and lower latitudes on earth are concentrated here. The Plateau is the source of many rivers in China and Asia. The Yangtze River and the Yellow River start from here, and flow across China's first, second and third steps. The Lantsang River (lower reaches of which is the Mekong River), the Nujiang River (lower reaches of which is the Salween River), the Shiquan River (lower reaches of which is the Indus River) and the Yarlung Tsangpo River (lower reaches of which is the Brahmaputra River) all origin from here. The Tarim River, the Ruo Shui River and the Shiyang River, flowing towards the Inner Mongolia-Xinjiang Plateau start from here as well. The Qinghai-Tibet Plateau is in the temperate zone. It is cold and dry. Despite little precipitation, the ice and snow can replenish the rivers and lakes. The water power resources are rich. The largest and highest lake group with the biggest number of lakes is situated here. Due to strong evaporation at low latitudes, most lakes become inland ones, only a few flow out. Thus, salt lakes and saltwater lakes dominate. The marshes on meadow steppes are located in or near the source region of rivers on the plateau, and develop in floodplains and low-lying lands. The peat boggy soil and the meadow boggy soil are the main soil types on wetlands. The most special marsh types are the *Kobresia tibetica* marsh and the *Carex muliensis* marsh. China's largest marsh peat is situated in Zoigê, northeast of the Plateau, under the cold and humid ground environment. Wetlands only have a few kinds and a small number of plankton and benthonic animals. Saline-alkaline tolerant fishes are largely distributed. A large number of waterfowls inhabit in the Qinghai Lake, the Gyaring Lake, the Ngoring Lake, the Namtso Lake, Zoigê and the middle reaches of the Yarlung Tsangpo River. The waterfowls have a large quantity but a few species, the main of which are bar-headed gooses, great black-headed gulls, brown-headed gulls and great cormorants. Zoigê is the only place in the world for the rare plateau crane—the black-necked crane to breed. The middle reaches of the Yarlung Tsangpo River is one of the major places for them to spend the winter.

Humans are sparsely populated here. Wetlands are seldom exploited, and most of them are in a natural state. But the conditions of wetlands here are closely connected with China and even Asia, and can have a far-reaching influence on them. The famous rivers, which start from here, have a long history in affecting the life of nearly two billion Asians. The tiny changes of the hydrology and climate on wetlands will spread to China and Asia as well. We should exert our best to protect "China's Water Tower" and "Asian Water Tower".

万山之父　湿地之母·西藏珠穆朗玛峰
在这以世界最高峰领衔的青藏高原上，众多的冰川雪山使高原的风光变得如此美丽。中国是中低纬度冰川发育最多的国家，冰川面积分别占世界和亚洲山地冰川总面积的 14.5% 和 47.6%，每年可融化水约 360 亿立方米。

Father of Mountains and Mother of Wetlands — the Mount Everest, Tibet
On the Qinghai-Tibet Plateau, where the highest mount in the world locates, glaciers and snow mountains add more beauty to the scenery. China has the most glaciers on the middle and lower latitudes. The area of glaciers in China accounts for 14.5% and 47.6% of the total area of the world and of Asian mountainous glaciers respectively, 36 billion cubic meters of water is melted from which every year.

湿地之源·西藏喜马拉雅山北坡
正因为有了这些冰川雪山，江河湖沼才有了原始的源泉，也才有了如此持久旺盛的生命，才有了江河横溢、万里迢迢流进大海的不竭动力。

Source of Wetlands — Northern Slope of the Himalaya Mountains, Tibet
The water melting from the glaciers and snow mountains becomes the primitive source of rivers, lakes and marshes, gives lasting vigor to life, and drives rivers to the ocean.

纳木错湖 · 西藏那曲

清澈而冰凉的纳木错湖水荡涤着念青唐古拉山，该湖湖面海拔高达 4718 米，是世界上最高海拔地区的最大湖泊，但它只是藏北高原上星罗棋布的数千个湖泊之一。青藏高原的湖泊面积占中国湖泊总面积的 25%，数量占到全国总数的一半，是世界上湖泊数量最多、湖面最高的湖区。这里的湖泊多为内陆湖，且多以盐湖和咸水湖为主。

The Namtso Lake — Nagqu, Tibet

The clean and cold water from the Namtso Lake cleanses the Nyenchen Tanglha Mountains. The Namtso Lake, with the elevation as high as 4,718 meters, is the world's largest lake in the highest elevation region. However, it is only one of the thousands of lakes scattered on the Northern Tibetan Plateau. The area of lakes on the Qinghai-Tibet Plateau accounts for 25% of China's total, and the number of lakes here is half of China's total,which makes it the lake region with the biggest number and highest altitude in the world highest and has the most lakes. Most lakes here are inland ones, dominated by salt lakes and saltwater lakes.

川西高山湖泊 · 四川理塘海子山

这里气候严寒而干旱，虽降水稀少，但冰雪融水是河流、湖泊的主要补给形式，水流在地势的低凹处或山口等集水区上形成了各种湖泊。在海子山这块平均海拔 4500 米，300 多平方千米大的山原面上，有着大小不一的 1140 多个海子（湖泊），犹如老天爷失手撒下的无数颗珍珠闪烁在山谷之间，其规模密度在中国是独一无二的。

Lakes on High Mountains in West Sichuan Province — Haizi Mountain, Litang County, Sichuan

The climate here is frigid and dry. Despite little precipitation, the melt water from ice and snow replenishes rivers and lakes. Various lakes are developed in the water-concentration region, low-lying places and mountain passes. The average elevation is 4,500 meters. On the mountain plateau, which is over 300 square kilometers, more than 1,140 lakes (*haizi*) with different sizes are situated, glittering between valleys like numerous pearls scattered by God by accident. The scale and density of lakes here is incomparable all over China.

繁星般的沼泽湖泊·青海黄河源区

在世界屋脊的高原面上，我们曾经弄不清黄河长江到底源自何处，面对遍地湖沼处处水，也只能发出黄河之水天上来的感慨。这里寒冷湿润的地面环境条件形成了中国面积最大的沼泽区，沼泽主要分布在高原的江河源区或者近河源区。

Starry Marshes and Lakes — Source Region of the Yellow River, Qinghai

On the plateau of "the roof of the world", we were confused where the Yellow River and the Yangtze River come from. Faced with lakes and marshes everywhere, we can only sigh that "the Yellow River comes from the sky". The cold and humid ground environment here shapes China's largest marsh region, with marshes mainly distributed in or near the source region of rivers on the plateau.

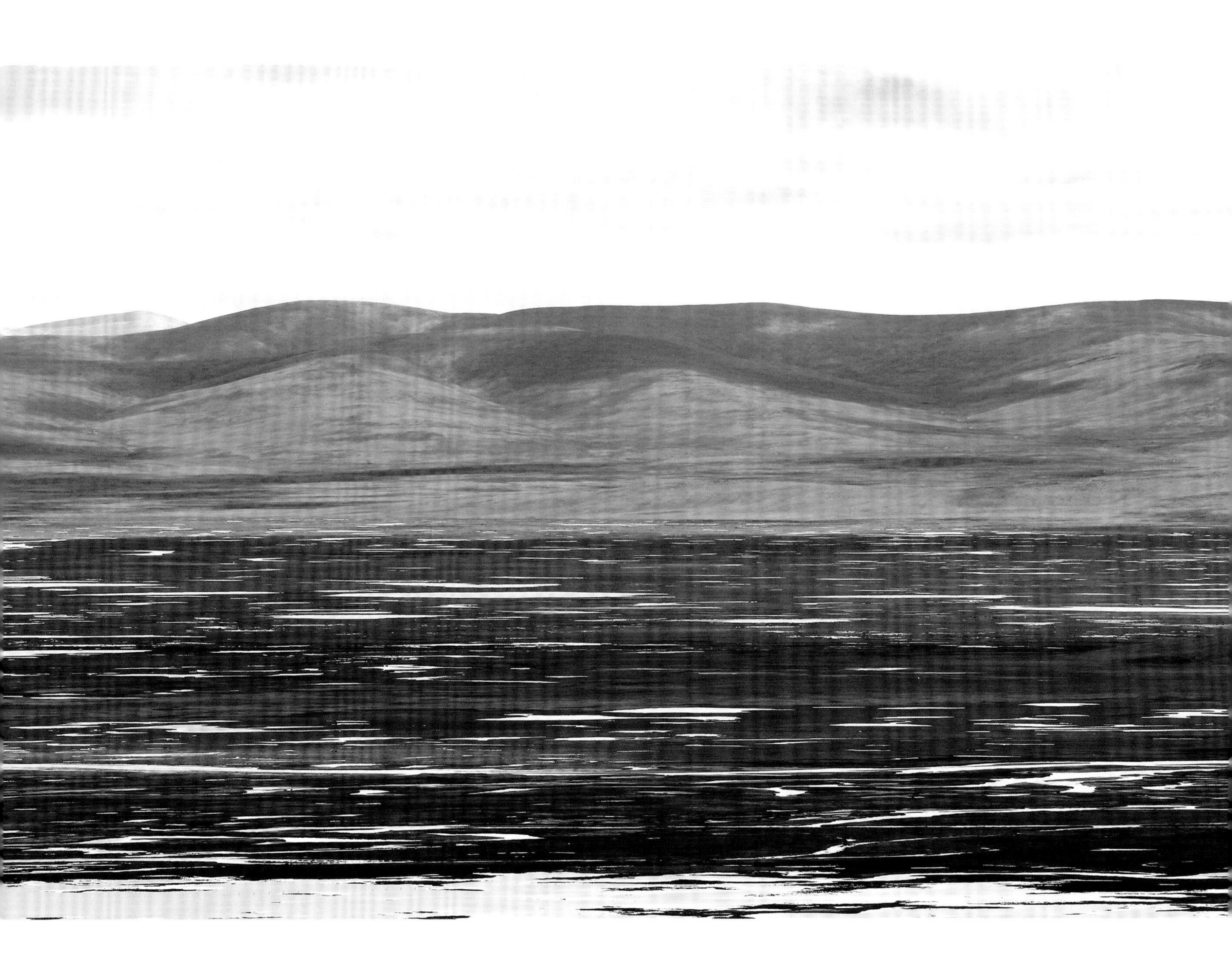

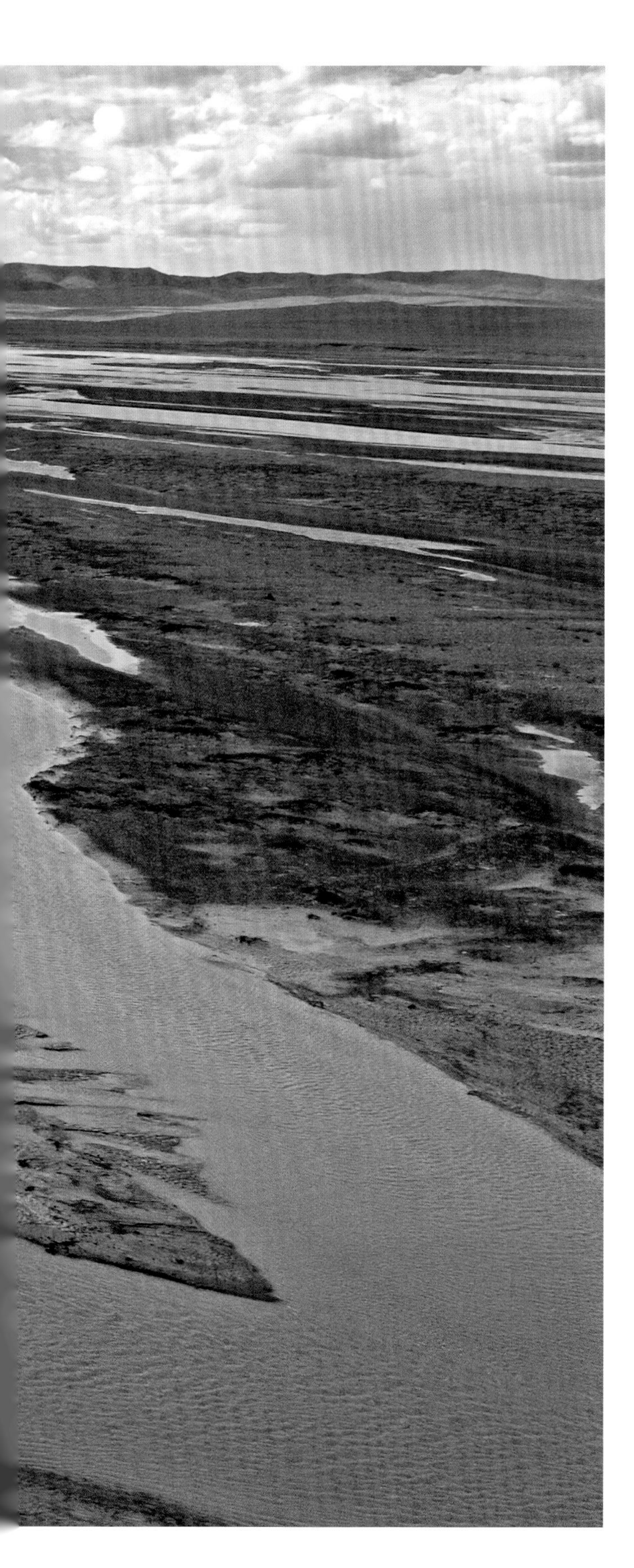

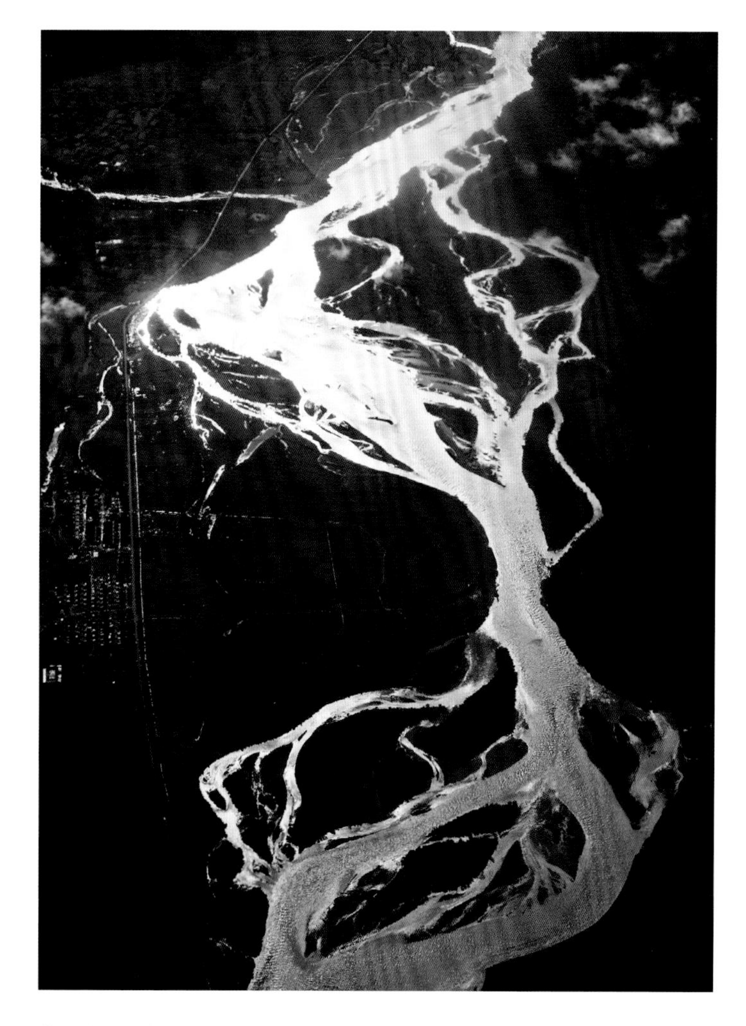

信马由缰 · 青海澜沧江源区

水在大地上随意流淌，无拘无束——自然的河流本来就是这样的。澜沧江发源青海，经西藏、云南出境后叫湄公河，流经缅甸、老挝、泰国、柬埔寨、越南五国后入南中国海，是亚洲流经国家最多的河流，称为“东方多瑙河”。

Flowing Freely — Headwaters of the Lantsang River, Qinghai

Water flows freely on land, with no restrains—exhibiting the natural state of rivers. The Lantsang River starts from Qinghai Province, passes through Tibet and leaves China from Yunnan Province. It is then called the Mekong River. It runs through Burma, Laos, Thailand, Cambodia and Vietnam, and finally flows into the South China Sea. As the river running through the most countries in Asia, the Mekong River is also known as "the Oriental Danube".

万里长江源头 · 青海海西

沱沱河为长江正源，发源于唐古拉山主峰的格拉丹东雪山的冰川中。长江全长 6397 千米，是中国第一、世界第三大河流，也是世界上最长的完全在一国境内的河流，流域面积 180 万平方千米，约占全国土地总面积的 1/5，与黄河一起并称为中华民族的“母亲河”。

Source of the Yangtze River — Haixi, Qinghai

As the source of the Yangtze River, the Ulan Moron (the Tuotuo River) begins in the glaciers of the Geladaindong Peak, the main peak of the Tanggula Mountains. The Yangtze River is 6,397 kilometres long. It is China's first and the world's third largest river, and also the longest river that flows within one country all over the world. The watershed area of it is 1.8 million square kilometres, accounting for 1/5 of the national land area together with the Yellow River, the Yangtze River is called the Mother River of Chinese nation.

山众曲多水富 · 青海囊谦

澜沧江上游水系的扎曲、孜曲、巴曲、热曲、吉曲五条河贯穿囊谦县全境，全县自产人均水量达 6.55 万立方米，为世界人均水量的 7.4 倍，为我国人均水量的 24 倍。这个数字，在青藏高原颇具代表性。

Mountainous Region Covered with Rivers, and Rich in Water Resources — Nangqên, Qinghai

At the upper reaches of the Lantsang River, five rivers, namely the Zhaqu River, the Ziqu River, the Baqu River, the Requ River and the Jiqu River run throughout Nangqên County. The county's per capita water generation is as high as 65,500 cubic metres, 7.4 times of that of the world average and 24 times of China's average. This number is very representative on the Qinghai-Tibet Plateau.

咆哮而下怒江水 · 西藏昌都

由于山高坡陡，河流趁势奔腾而下，波涛汹涌怒气冲天，怒江因此得名。怒江发源于西藏的唐古拉山南麓，经云南流入缅甸后改称萨尔温江，最后注入印度洋。

The Roaring Nu River — Changdu, Tibet

From high mountains and steep slopes, the Nu River roars down, with anger and rage, thus the name "Angry River" ("Nu" names angry in Chinese) is given. Beginning from the south Tanggula Mountains, Tibet, the Nu River passes Yunnan Province, enters Burma, and is then called the Salween River. At last, it flows into the Indian Ocean.

沙滩的舞台　雪水的画 · 西藏日喀则

雅鲁藏布江是西藏最大的河流，在藏语中意为"高山流下的雪水"，中游宽广的河漫滩上雪水任意流淌着。该江出境后称之为布拉马普特拉河，经南亚次大陆的印度、孟加拉国后注入印度洋。

Arena of Sands and Picture of Snow Mountains — Shigatse, Tibet

The Yarlung Tsangpo River is the largest river in Tibet. Its name means "the snow water from high mountains" in Tibetan. The snow water flows freely in the wide middle reaches of the river. After leaving China, the river is called the Brahmaputra River. Passing India and Bangladesh on the South-Asian sub-continent, it flows into the Indian Ocean finally.

雪山冻湖冰花开·青海海北

湖面美丽的冰花是由湖底沉积生物产生的气体所成，远处是东西走向积雪的祁连山。祁连山南坡的水由石羊河、黑河、疏勒河等河流输向河西走廊，并最终在内蒙古沙漠中消失，为内流河。

Under Snow Mountains, Ice Flowers Blossoming on Frozen Lakes — Haibei, Qinghai

The beautiful ice flowers on the lake surface are formed by gases released by sediments on the lakebed. In the distance, lie the Qilian Mountains covered with snow. Water from the southern slope of the Mountains passes the Shiyang River, the Heihe River and the Shule River, etc., runs towards the Hexi Corridor, and finally, disappears on the desert in Inner Mongolia. It is an interior river.

肥厚的泥炭地 · 四川若尔盖

在水流不断冲刷下形成的沟壁上可以看到泥炭层的剖面。若尔盖湿地是中国最大的泥炭沼泽湿地，湿地面积 80 万公顷，储存的泥炭高达 19 亿吨，泥炭层可厚达 10 米多。泥炭能储藏自身重量 4 ~ 10 倍的水，在枯水季节，黄河经过这里所增加的径流量可达 45% 之多。可是，近年来由于气候变化和人为开垦、过度放牧等原因，沙化土地正以每年近 10% 的速度增长，严重威胁着这片珍贵的湿地。

Fertile Peat Land — Zoigê, Sichuan

On the trench wall eroded by water, the profile of the peat is exposed. Zoigê wetland is China's largest peat marsh wetland. The peat is over 10 metres thick and can store water weighing 4 to 10 times of the peat itself. In dry seasons, the runoff volume of the Yellow River increases by as much as 45%. However, in recent years, due to climate change, reclamation and over grazing, the desertified land is widening by nearly 10% annually, threatening the precious wetland seriously.

1

2

3

4

1. 湖畔泥沼 · 西藏那曲
Lakeside Swamp — Naqu, Tibet

3. 苔草沼泽 · 西藏定日
Carex Marsh — Tingri, Tibet

2. 沼泽与沙棘林 · 西藏曲水
Marsh and Sea-buckthorn Forest — Qushui, Tibet

4. 湿地中的赤麻鸭和黑颈鹤 · 青海玛多
Ruddy Shelducks and Black-necked Cranes on Wetlands — Madoi, Qinghai

沼泽地上的牦牛群 · 青海玉树

雪地、沼泽、牦牛群是青藏高原最具代表性的生态画面。但是，目前青藏高原的高寒草甸（含沼泽）正在以每年 50 平方千米的速度消失，荒漠化扩大的速度在加剧，出现了较为严重的生态退化问题。

Yaks on Marshes — Yushu, Qinghai

Snowfields, marshes and yaks can best represent the Qinghai-Tibet Plateau. However, the Plateau's alpine meadow (marshes included) is disappearing by 50 square kilometers per year. With the rate of desertification aggravated, serious ecological degradation is occuring.

秋色中的藏野驴 · 青海黄河源区

红色的碱蓬、黄色的苔草相互交织着，仿佛一幅绚丽秋色的地毯。一群藏野驴却无心顾及周边美丽的景色，而是抓紧时间进食抓膘，准备度过即将到来的艰难漫长的冬季。

Kiangs in Autumn — Headwaters of the Yellow River, Qinghai

Red suaeda glauca and yellow carex are interwoven into a gorgeous carpet of autumn. Paying no attention to the beautiful scenery, a group of kiangs hurry up to feed themselves, preparing for the upcoming rough winter.

难得一见的集群 · 四川若尔盖月亮湾

有别于越冬地，在繁殖地黑颈鹤的大量集群是难得一见的，聚集的黑颈鹤竟然有175只之多。只有在迁飞南方越冬之前，散布于各处的一家家黑颈鹤才会带着出生不久的雏鸟，来到月亮湾短暂集聚后准备飞到南边的云贵高原越冬，它们要依靠团队的力量来共同战胜长途迁徙中的万般艰辛。

Group Hardly Seen — the Moon Bay, Zoigê, Sichuan

Unlike where they spend the winter, in their breeding place, a large group of black-necked cranes can hardly be seen. There are as many as 175 black-necked cranes. Only before migrating to the south to spend the winter, will the scattered crane families take their new-born baby, gather shortly at the Moon Bay and fly to the Yunnan-Guizhou Plateau in the south. They need to rely on the group to teams to conquer hardship during the long journey.

黑颈鹤 · 四川若尔盖

这里是珍稀高原鹤类——黑颈鹤的主要繁殖地。黑颈鹤是世界 15 种鹤中唯一一种栖息在高原地区的鹤。它在青藏高原北部及临近地区海拔 3500 ~ 4500 米的沼泽繁殖，在青藏高原南部和云贵高原海拔 2500 ~ 3500 米的高原湿地过冬。仅有少数个体在印度越冬。

Black-necked Cranes — Zoigê, Sichuan

This is the breeding place for black-necked cranes—a rare and precious kind of plateau cranes. Of all 15 species of cranes in the world, the blacked-necked crane is the only one that lives in plateau regions. They breed in swamps in the northern regions of the Qinghai-Tibet Plateau, with an altitude of 3,500 to 4,500 meters. Their wintering grounds are wetlands on the southern Qinghai-Tibet Plateau and the Yunnan-Guizhou Plateau, with an altitude of 2,500 to 3,500 meters. Only a small number of individuals choose India to spend the winter.

棕头鸥群起飞 · 青海湖

青海湖是中国湖泊中面积最大的湖，同时也是我国第一大咸水湖。鸟岛上栖息着数以万计的棕头鸥、斑头雁、普通鸬鹚、渔鸥等，候鸟数量甚多但种类较少。在繁殖季节，各式各样的鸟巢密密麻麻，五颜六色的鸟蛋遍地皆是，群鸟翩然飞翔，遮天蔽日，甚为壮观。青海湖是中国的首批国际重要湿地之一。

Brown-headed Gulls Taking off — Qinghai Lake

The Qinghai Lake is China's largest lake and largest saltwater lake as well. On the Bird Island, inhabit tens of thousands of brown-headed gulls, black-headed gooses, great cormorants and pallas's gulls, etc. Migrant birds are large in quantity, but few in species. In breeding seasons, various kinds of bird nests are thickly dotted, and colourful bird eggs can be found everywhere. Groups of birds fly up in the sky, like a black cloud, drawing a spectacular picture. The Qinghai Lake is one of China's first group of wetlands of international importance.

初沐阳光 · 青海湖
小斑头雁孵化出壳了，初升的太阳照在雏鸟身上暖意融融。新生命的世界多么美好。

Born in the Sun — Qinghai Lake
Little black-headed gooses are hatched, bathing in the warm sunshine. What a beautiful world it is for a new life!

神湖的祈祷 · 西藏纳木错

在藏族人民心目中神圣不可侵犯的湖有不少，纳木错就是其中最著名的神湖之一。僧俗信众不惜长途跋涉，到这里转湖念经，得到的是莫大的安慰和幸福，这是千百年来在藏区一直延续的文化传统。

Prayer of the Sacred Lake — Namtso, Tibet

For the Tibetans, there are many sacred lakes. Namtso is one of the most important ones. Monks, laymen and believers travel a long distance to chant scriptures around the lake, in the hope of getting solace and blessings. It is a cultural tradition among the Tibetans for centuries.

"阳光照耀着塔什库尔干"· 新疆塔什库尔干

这里是中国地势第一阶梯最西端的帕米尔高原，也是中国国土的最西端。塔吉克族的小姑娘在湿草地上放牧。

"Tashkurgan under the Sun" — Tashkurgan, Xinjiang

This is the westernmost of the first step of China's terrain, the Pamir Mountains; and is also the west end of China's territory. The Tajik girls herd cattle on wet grasslands.

美丽的传说　神奇的湖 · 青海多卡寺

相传鄂陵湖畔的多卡寺是松赞干布为迎娶文成公主曾经等了近两个月的地方。鄂陵湖和扎陵湖是黄河源头两个最大的高原淡水湖泊，湖面海拔 4300 多米，两湖非常接近，面积也相差不大，素有“黄河源头情侣湖”之称。

Beautiful Legend and Magical Lake — Duoka Temple, Qinghai

Legend says the Duoka Temple by the Ngoring Lake is where King Songtsan Gampo waited for almost two months when he married Princess Wencheng, daughter of an emperor in Tang Dynasty. The Ngoring Lake and the Gyaring Lake are two largest freshwater lakes on the plateau at the source of the Yellow River. The elevation of the lakes' surface is over 4,300 meters. The two lakes are close to each other, and are almost of the same size, thus are called "the couple lakes at the source of the Yellow River".

2 蒙新高原湿地区

Inner Mongolia-Xinjiang Plateau Wetland Zone

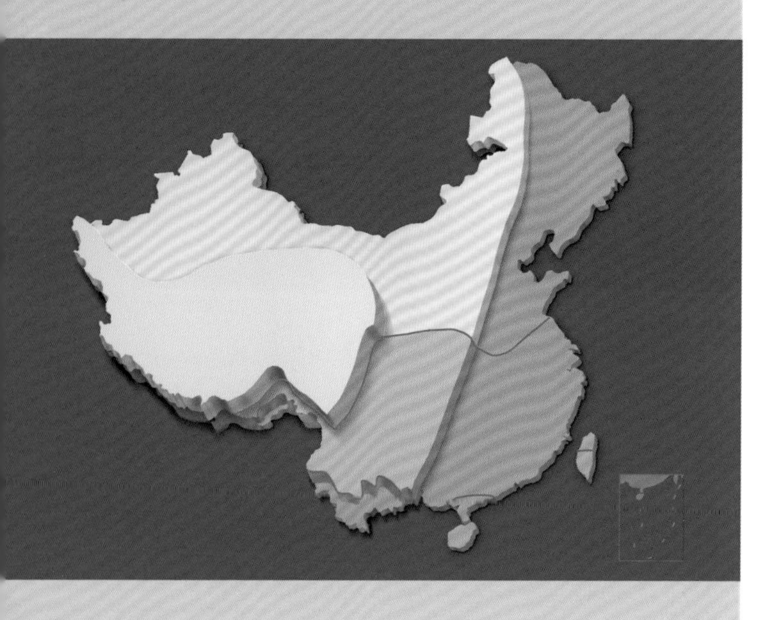

蒙新高原湿地区

我国最北部的蒙新高原湿地区包括我国青藏高原以北的新疆、甘肃，秦岭以北包括内蒙古、宁夏和黄土高原的所有地区，东至大兴安岭、太行山以西，属中国地势第二级阶梯的北部分，气候区属温带。黄河为该地区最大的过境河流。除大兴安岭北段湿度较大以外，其余大部分地区的特点是地处内陆腹地，气候干旱，降水稀少，蒸发量大，地表径流补给不丰，除少数如额尔齐斯河、喀纳斯湖、乌梁素海等为外流河（湖）外，多为内流河（湖），如塔里木河、黑河、疏勒河、博斯腾湖等。湖水因不断被浓缩而发育成闭流类的咸水湖或盐湖。该地区沙漠广袤，在沙漠边缘沙丘洼地有一些中小湖泊分布，为本区域湖泊的又一大特色。在寒冷干燥的环境条件下，仅有局部水源补给十分丰富的地方才有少量沼泽发育，盐碱沼泽中耐盐性植物成分较多，以嵩草、苔草、芦苇等草本为主，木本、藓类沼泽不发育。浮游生物、底栖动物种类少、数量低，生长有一些耐盐碱的鱼类，种类也较单一。该地区因其自然隐蔽条件好，以及部分短途和横向迁徙的水禽自然习惯性因素的作用，每年有众多的水禽在此栖息繁殖，如大天鹅、斑头雁、赤麻鸭、黑鹳、棕头鸥、红嘴鸥、红脚鹬等。另外，塔里木河流域是我国黑鹳的重要繁殖地。内蒙古毛乌素沙地中的湖泊群承载了自然界遗鸥总数40%以上的个体，为遗鸥最重要的繁殖地。新疆布尔根河是我国河狸的唯一栖息地。

这里的湿地生态系统相当脆弱，人与湿地之间的关系也异常敏感，由于蒸发量大大超过降水量的地区气候特点，无论是降水、来水还是过境水，水量的适时补充是维系这里湿地存在的生命线，也是维系人们生存的荒漠绿洲的生命线。有水就有生命、有水才有绿色，可持续的保护及合理利用湿地，是人与湿地能否和谐相处的关键，更是影响人们能否在这里生存下去的最直接原因。

（前页）秋天的草原　蜿蜒的河 · 内蒙古大兴安岭

河流在高原面上蜿蜒流过，划出了一个漂亮的蓝色"S"形，周边是草原、灌丛和零散树木。这里是中国纬度最高的地区，绚丽的秋色告诉人们，长达半年河湖封冻、冰天雪地的冬季即将来临。

(Previous Page) Grasslands and Winding Rivers in autumn — Greater Khingan Range, Inner Mongolia

Rivers wind on the plateau, leaving a beautiful blue "S" shape. Grasslands, shrubs and scattered trees can be seen on riverbanks. This place has China's highest latitude. The gorgeous autumn tells people that the half-a-year-long winter of ice and snow with freezing rivers and lakes is about to come.

（从左至右）
河滩冬雪 · 新疆布尔津
大青山下的哈素海 · 内蒙古呼和浩特
草原沼泽 · 内蒙古呼伦贝尔
湿地中的黑鹳 · 新疆塔里木河
盐场鸟瞰 · 宁夏毛乌素沙地
(From Left to Right)
Snow on Riverbanks — Burqin, Xinjiang
Lake Hasuhai under Daqing Mountain — Hohhot, Inner Mongolia
Swamps on Grasslands — Hulun Buir, Inner Mongolia
Black Storks on Wetlands — Tarim River, Xinjiang
An Aerial View of Saltern — Mu Us Desert, Ningxia

山前冲积扇 · 内蒙古阴山
雪水和降雨变为溪流向下流淌，当山地河水流至山麓出口进入开阔平坦地区后，由于河床坡度变缓，流速减小，水流呈放射状向外流动，就形成了上窄下宽的冲积扇。蒙新高原分布有很多大面积的山前冲积扇，冲积扇是本区域湿地水源的主要渠道之一，也是大部分荒漠绿洲产生的重要前提。

Alluvial Fan in front of Mountains — the Yin Mountains, Inner Mongolia
Snow water and rainfall turn into streams, flowing down. When rivers on top flow to the foot of mountains, and stretch to the wide and flat region, the slope of riverbeds flattens, the flow rate slows down, and waters radiate, becoming an "A"-shaped alluvial fan. Many large alluvial fans in front of mountains are distributed in the Inner Mongolia-Xinjiang Plateau. Alluvial fans are a major water source of wetlands in this region, and an important condition for the development of most desert oases.

沙与水的辩证法 · 内蒙古赤峰
在这里，沙与水是相辅相成的，没有沙丘就没有丘间水，沙水共生而非相克，充满了自然界对立统一的辩证关系。

Dialectical Relationship between Sand and Water — Chifeng, Inner Mongolia
Sand and water complement each other here. Without sand dunes, there will be no water between dunes. Sand and water reinforce rather than restrict each other, forming a dialectical relationship of unity of opposites in the natural world.

黄河之水天上来·山西壶口

黄河从青藏高原一路下来，途经甘肃、宁夏、内蒙古及黄土高原，携带着大量的泥沙咆哮而过。黄河是本区域中最大的过境水，它使本区域，尤其是前后河套地区得到灌溉和舟楫之利的同时，也使黄土高原的水土大量流失，下游河道淤积阻塞、灾害频发。黄河是中国北方地区最大的供水水源，它以占全国河川 2% 的有限水量，灌溉着近 20 万平方千米的耕地，承担着中国 1.2 亿人口和 50 多座大中城市的供水压力，已经不堪重负。

The Yellow River Flowing from the Sky — Hukou Waterfall, Shanxi

The Yellow River flows down from the Qinghai-Tibet Plateau, and passes Gansu, Ningxia, Inner Mongolia and the Loess Plateau, roaring and carrying a large amount of sediments. The Yellow River is the largest transiting water in this region. On the one hand, it benefits the region, especially the Hetao region, with irrigation, and provides

water channels for ships and boats. On the other hand, it also leads to soil and water erosion on the Loess Plateau, river channels sedimentation at the lower reaches and frequent disasters. The Yellow River is the largest water supply in the north of China. Accounting for 2% of the water yield in China's rivers, the Yellow River irrigates farmlands covering nearly 200,000 square kilometers, and burdens itself with water supply for 120 million Chinese people and over 50 large-and-medium cities. It can hardly bear the load.

泥炭沼泽及浮水植物 · 内蒙古大兴安岭

大兴安岭北段西坡的内蒙古高原面上，由于特殊的地理气候条件，也有沼泽生成。这是秋天枯水季节时的泥炭沼泽。

Peat Bogs and Floating Plants — Greater Khingan Range, Inner Mongolia

In the Inner Mongolian Plateau on the western slope of the north Greater Khingan Range, with special geographical and climatic conditions, swamps are also developed. This is the peat bog in the dry season of autumn.

秋色卧龙湾 · 新疆喀纳斯

喀纳斯湖是额尔齐斯河最大支流布尔津河的发源地，而额尔齐斯河水系是中国唯一的北冰洋水系，其气温、降水等都主要受北冰洋影响，该区域是我国唯一的南西伯利亚区系动植物的分布区，北湿南干是这里的自然气候特点。

Wolong Bay in Autumn — Kanas, Xinjiang

The Kanas Lake is the source of the Burkin River, the largest tributary of the Irtysh River. The Irtysh River system is China's only water system in the Arctic Ocean. The river's temperature and precipitation are affected by the Arctic Ocean. The region is China's only southern Siberian fauna and flora distribution area. The climate characteristic here is humid in the north and dry in the south.

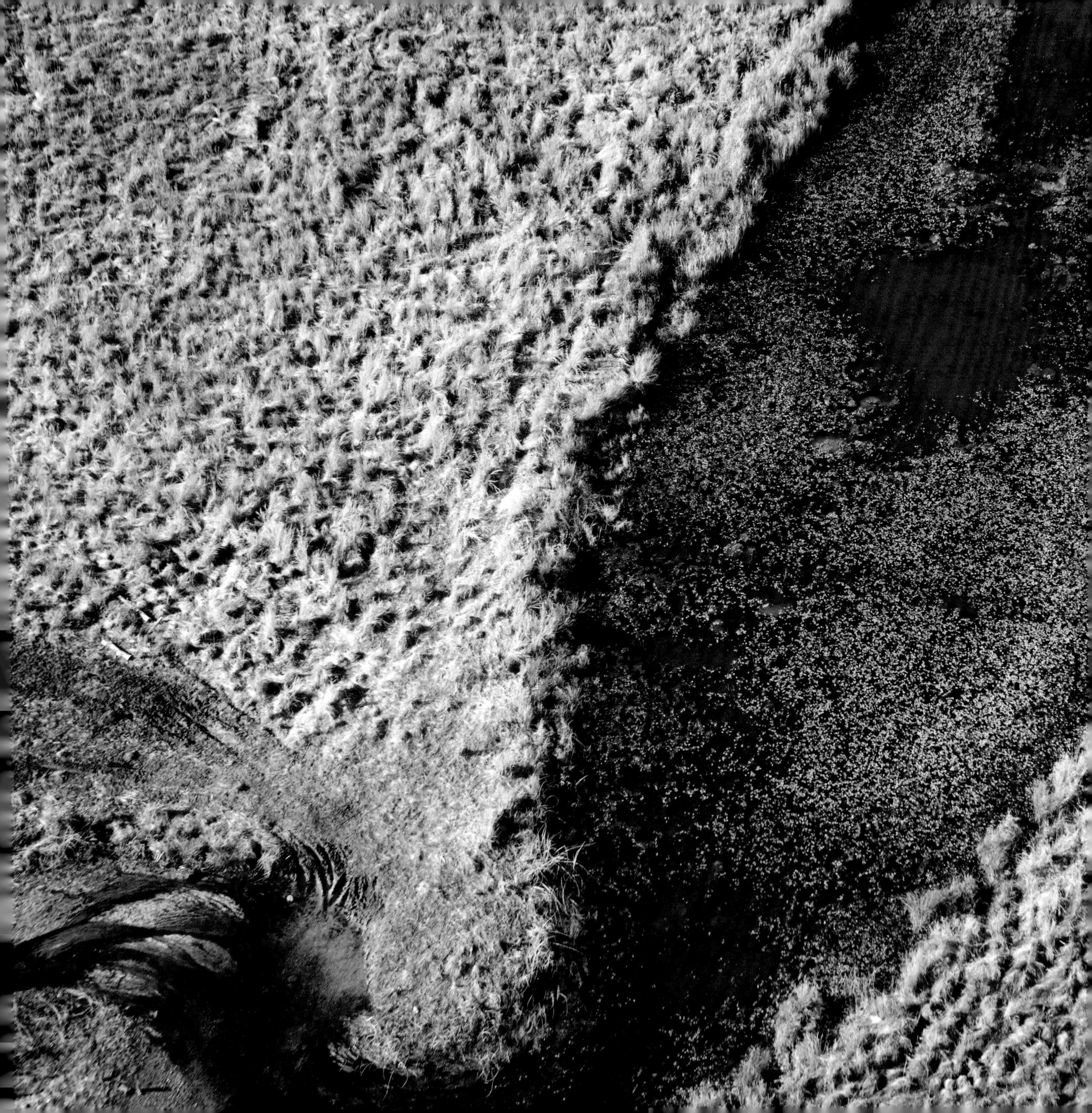

塔里木河胡杨 · 新疆塔里木河

塔里木河为中国最大、世界第五大内流河，它是保障塔里木盆地自然生态、绿洲经济和各族人民生活最重要的"生命之河"，塔里木河流域的胡杨林是世界上面积最大的胡杨林。由于上游来水减少，致使河段断流，大片胡杨树枯死，沟通塔里木盆地南北的绿色走廊将面临毁灭的危险。

Euphrates Poplars by Tarim River — Tarim River, Xinjiang

As China's largest and the world's fifth largest interior river, the Tarim River maintains nature and ecology of the Tarim Basin, the oasis's economy, and is the most important "river of life" for people of various ethnic groups here. The Euphrates poplar forest in the Tarim River Basin is the largest in the world. As water from the upper reaches decrease, rivers are cut off, a vast expanse of euphrates poplar forests withers, and the green corridor linking the north and the south Tarim Basin faces the risk of destruction.

湿地牧场 · 新疆阿勒泰

这里是中国版图的最西北，主要受北冰洋气候影响，山谷中湿润河畔的沼泽草甸水草肥美，是极好的湿地牧场。

Wetland Pasture — Altay, Xinjiang

This place lies on the northwestern edge of China's territory. Influenced mainly by the Arctic Ocean climate, the meadow is fertile on the riverside swamps of the valley, and is a wonderful wetland pasture.

析出 · 内蒙古锡林郭勒

蒙新高原的很多湖泊因干旱少雨、蒸发量大等气候因素，不断被浓缩发育成半咸水湖、咸水湖或盐湖，盐湖湖水的盐浓度到达一定程度后，盐就会自然析出。

Precipitation — Xilingol, Inner Mongolia

Due to drought, little rainfall, high evaporation and other climatic elements, many lakes on the Inner Mongolia-Xinjiang Plateau are constantly concentrated and develop into brackish-water lakes, salt water lakes or salt lakes. As the salinity in the salt lakes reaches to a certain extent, salt can be naturally precipitated.

沙漠深处天鹅湖 · 内蒙古阿拉善

这是腾格里沙漠腹地的天鹅湖，人迹罕至，仍呈原始状态，有天鹅、野鸭、鹬鹰等水鸟在这里自由自在地生活。在沙漠或沙地的丘间洼地里分布着的一些中小型湖泊，往往是很多鸟类极好的隐蔽点和栖息地，这是本区域湖泊的一大特色。近年来由于旅游开发，这个景象已经彻底消失。

Swan Lake in Deep Deserts — Alxa, Inner Mongolia

This is the swan lake in the deep Tengger Desert. Few people can be seen here. The lake is still in its virgin state. Swans, wild ducks and other waterfowls live at freewill. Distributed on sand dunes and low-lying lands in the deserts, small-and-medium-sized lakes are wonderful shelters and habitats for many birds. This is a characteristic for lakes here. In recent years, due to tourism development, such scenery has completely disappeared.

林下塔头沼泽 · 新疆阿勒泰

塔头沼泽是一种主要由无数高出水面几十厘米甚至一米多的草墩组成的沼泽，由沼泽地里各种苔草死亡后再生长、再腐烂，周而复始，并和泥炭长年累月凝结而成。林下塔头沼泽在本区域的新疆林区及内蒙古北部林区有分布，森林树种在新疆主要为雪岭云杉、西伯利亚落叶松，在内蒙古北部主要为落叶松、云杉和冷杉。

Undergrowth Grassy Marshland — Altay, Xinjiang

The grassy marshland is a marsh, made up of blocks of grass, dozens of centimetres or over 1 meter higher than the water surface. On the marshland, various kinds of carex die, regrow and decay over and over again. After months and years, under the effect of peat, the grassy marshland is developed. The undergrowth grassy marshlands are distributed in forests in Xinjiang and northern Inner Mongolia. Typical tree species in Xinjiang are *Picea schrenkiana* and Siberian larches; in northern Inner Mongolia, larches, spruces and firs.

夕阳黑河 · 内蒙古额济纳

黑河是我国西北地区仅次于塔里木河的第二大内陆河，发源于祁连山，终结于内蒙古居延海。上中游用水量的增大曾使得下游断流，居延海一度干涸。

Ruo Shui River under Sunset — Ejin, Inner Mongolia

The Ruo Shui River is largest interior river in the northwest part of China, only second to the Tarim River. It starts from the Qilian Mountains, and ends in the Juyan Lake, Inner Mongolia. The water use increase in the upper reaches once cut off the lower reaches, leaving the Juyan Lake dried up for a while.

草原湿地 · 内蒙古苏尼特

蒙古语"淖尔"意为"湖泊"，也叫"水泡子"，往往为草原洼地积水而成，有的是内流河的蓄水部分或河尾。内蒙古草原上分布有很多大大小小这样的湖泊，有些小的湖泊雨季出现、旱季消失。它们是草原生态系统的重要组成部分，是维系草原湿地生物不可或缺的生命线。

Grassland Wetlands — Sunite, Inner Mongolia

"Naoer" in Mongolia means "lakes" or "ponds", which were normally formed in the depression of grassland as water flows in. Some of which are water storage or the end of an inland river. Numerous such lakes of various sizes are distributed in the grassland of Inner Mongolia, some small ones are seasonal: emerge in raining season and disappear in dry season. They are important component of the grassland ecosystem and indispensable life line for maintaining grassland wetland beings.

中国最低的湿地 · 新疆吐鲁番

艾丁湖湖面海拔 -154 米，是全国最低的洼地、湿地，也是世界上主要洼地之一。艾丁湖水，主要来自西部和北部山脉的冰川水汇流，以及盆地北缘涌出的天山雪水潜流。由于冰川退缩、湖水蒸发、上游工农业用水的截留以及机井抽取使地下水位下降等原因，大部分湖面已变为深厚的盐层，只剩下很小的水面了。

Lowest Wetlands in China — Turpan, Xinjiang

With an elevation of -154 meters, the Ayding Lake is China's lowest low-lying land and wetland, and also one of the major low-lying lands in the world. Water of the Ayding Lake comes from the confluence of glacier water from the western and northern mountains and underflows of snow water gushing from the northern edge of the basin. Because of glacial recession, lake evaporation, cut-off of the water for industrial and agricultural usages along the upper reaches, and groundwater recession due to motor-pumped well drilling, most of the lake surface has become thick salt bed, leaving only a small water surface.

河狸筑起的水坝 · 新疆青河

河狸有一个独特本领是垒坝，当进入新的栖息地或栖息地水位下降时，河狸就会用树枝、泥巴等筑坝蓄水形成池塘和湿地，以保护洞口位于水下，防止天敌侵扰。因此河狸有动物世界“建筑师”之称。

Dams Built by Beavers — Qinghe, Xinjiang

One of the unique abilities of beavers is building dams. When moving to a new habitat, or when their habitat's water level decreases, beavers will build dams with branches and mud to store water. They will build ponds or wetlands to keep their caves below the water surface, and away from their enemies. sSo, they are called "architects" in the animal world.

晨曦中的河狸 · 新疆青河

新疆乌伦古河及上游的青格里河、布尔根河、查干郭勒河两岸，尤其是布尔根河流域生活着我国唯一的河狸种群。冬天即将来临，河狸忙着收集并存储自己爱吃的嫩树枝条，这个拖着树枝的夜行动物，忙得竟然没有顾及到清晨的天色已经开始发亮了。

Beavers at Dawn — Qinghe, Xinjiang

On banks of the Ulungur River and the Qingil River and the Chaganguole River at the upper reaches, the Burgen River in particular, live the only beaver groups in China. Winter is around the corner. Beavers are busy collecting and storing their favorite branches. This nocturnal animal, dragging branches, is so busy that it does not notice the sky is brightening up.

遗鸥飞来 · 陕西红碱淖

遗鸥几乎是最晚被发现的新鸥种，因此得名。作为繁殖地的毛乌素沙地湖泊，承载了自然界遗鸥总数的 40% 以上。遗鸥是为数不多的横向迁徙的鸟类，近年来发现，其越冬地位于渤海湾天津段一带。

Flying Relict Gulls — Hongjian Mire, Shaanxi Province

Relict gulls are almost the latest discovered gulls. This is also where their name comes from. As their breeding place, the Mu Us Desert Lake inhabits more than 40% of relict gulls in the natural world. Relict gulls are one of few kinds of birds that migrate crosswise. It is discovered in recent years that the wintering ground of relict gulls is the Tianjin section of the Bohai Bay.

水边的蓑羽鹤 · 内蒙古呼伦贝尔

蓑羽鹤是世界现存 15 种鹤中体型最小的一种，在中国分布较广，栖息于开阔草甸沼泽、芦苇沼泽和湖泊草甸等各类生境中，有时也到农田地活动。蓑羽鹤的主要越冬地在南亚和东非，它们有集群飞越喜马拉雅山的能力。

Demoiselle Cranes by Water — Hulun Buir, Inner Mongolia

Demoiselle cranes are the smallest cranes of the 15 crane species existing in the world today. They are widely distributed in China, and inhabit in open grassy marshlands, reed swamps, lake meadows and various other ecosystems, and sometimes in farmlands. The main wintering grounds for demoiselle cranes are South Asia and East Africa. They are also able to fly over the Himalaya Mountains in groups.

搏 · 新疆巴音布鲁克

巴音布鲁克是我国最大的天鹅繁衍、栖息地，每年 3 月到 4 月，以大天鹅、小天鹅、疣鼻天鹅为主的一万多只珍禽便飞到这里繁衍生息，直到 10 月、11 月离开，居留期长达半年以上。1980 年，巴音布鲁克建立了中国的第一个天鹅自然保护区。

Fighting — Bayanbulak, Xinjiang

Bayanbulak is the largest breeding habitat of swans in China. From March to April of each year, over 10 thousand rare birds dominated by Whooper Swans (*Cygnus cygnus*), Tundra Swans (*Cygnus columbianus*) and Mute Swans (*Cygnus olor*) would fly here to breed and inhabit until October and November. The residing period is longer than half a year. In 1980, the first Swan Nature Reserve of China was established in Bayinbrook.

迎着霞光·甘肃甘南

灰鹤多栖息于水洼地的沼泽草甸中，杂食性，但以植物为主。在中国的繁殖地主要就在本区域，灰鹤是世界上鹤类中分布最广的鹤种，广泛分布于欧亚大陆及非洲北部。

Flying towards Sunshine — Gannan, Gansu

Common cranes mainly inhabit on marshy meadows on low-lying lands. They are omnivores, but primarily feed on plants. Their major breeding place in China is here. Common cranes are the most broadly distributed cranes in the world. They are widely distributed on Eurasia and northern Africa.

转场·新疆喀纳斯

世世代代生活在布尔津河流域的哈萨克人，有“迁徙”在夏、冬草场之间的传统习惯，实质就是寒冷气候条件下的逐水草而居。

Migration between Grasslands — Kanas, Xinjiang

The Kazakhs living in the Burqin River Basin generations after generations has a traditional habit of "migrating" between summer and winter grasslands. In other words, they live where there is water and grass in the cold climate.

人进沙退还是沙进人退？·内蒙古巴彦淖尔黄河边

有水就有生命，有水就有绿色，人与沙的博弈武器就是水，也只能依靠水。在干旱的西北地区生存和发展，水比什么都重要。

Advancing or Retreating? — Bayannur by the Yeliow Rivier, Inner Mongolia

Where there is water, there is life. With water, green can be seen. The combating weapon between man and sand is water and nothing else. To live and thrive in the northwest, water is the most important of all.

3 云贵高原湿地区

Yunnan-Guizhou Plateau Wetland Zone

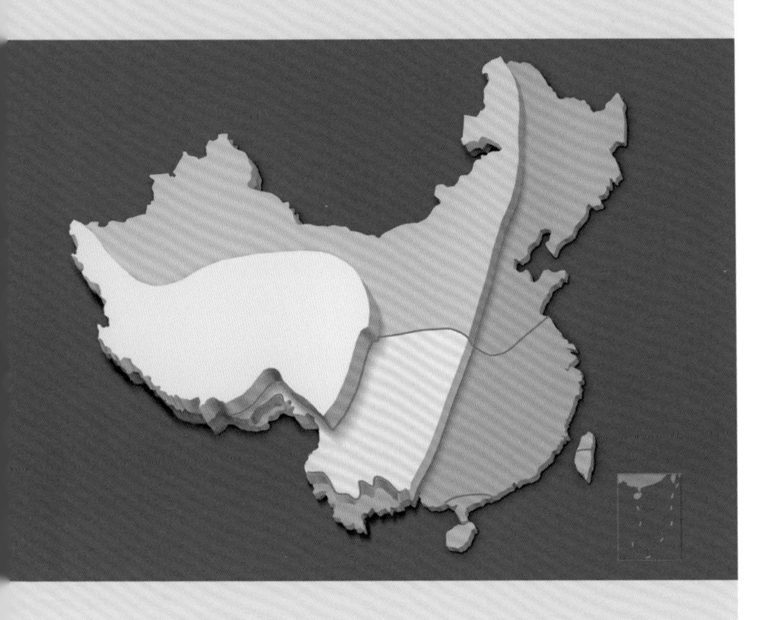

云贵高原湿地区

中国西南部的云贵高原湿地区北以秦岭为界，西为青藏高原，东至巫山、雪峰山一线，南至我国热带北缘，含云贵高原、四川盆地和西藏东南部，属中国地势的第二级阶梯，气候区为亚热带。虽属亚热带季风气候，但因低纬度、高海拔的原因，冬夏冷暖干湿差别不大，气候温和是其特点。该地区承接了来自第一阶梯的大河如长江、澜沧江、怒江等著名大江。由于西南季风带来的降水量大和地势的陡变，也发源了诸如中国第三大河流的珠江，长江的一级支流汉江、乌江等，以及元江（下游为红河）等大江和国际河流。云贵高原因为构造运动强烈，高山峡谷、小平坝和盆地等交错分布，一些湖泊往往都分布于断裂带或各大水系的分水岭地带，入湖支流水系较多，出水水系较少，湖泊换水周期长，生态系统较脆弱，虽然大都为淡水湖泊，但矿化度明显超过东部平原的湖泊。受西南季风影响，夏季降水往往占全年降水80%以上。湖泊水位随季节变化相差大，冬季无冰情，既有营养型湖泊也有贫营养型湖泊。该地区特有生物物种有如螺蛳、海菜花、金线鲃、裂腹鱼等。由于气候温和，发育的沼泽主要为芦苇沼泽和苔草沼泽，泥炭藓沼泽甚少，有如腾冲的北海等有分布。滇东北的大山包、滇西北的纳帕海及黔西北的草海等地是黑颈鹤的重要越冬地，昆明滇地附近是红嘴鸥的主要越冬地。

这里过境水很多，承上启下的地形地貌蕴藏了异常丰富的水利资源，“西电东送”、“南水北调”等大型工程是人们对这里湿地改造的重头戏，而环境影响评估却往往滞后和无力，这里湿地改造之后造成的影响是长远的和异常重大的，这些湿地改造工程将极大地影响着中国的能源、水分布及生态格局，利弊体大。因此，国家层面上的科学布局、环境影响评价、战略规划尤为重要，我们不能有丝毫的懈怠和忽视。

（前页）金沙江第一湾 · 云南迪庆
从第一阶梯下来的长江，进入了地势陡变的云贵高原高山峡谷区，山回水转。这是长江（上游称金沙江）进入本区域的第一个马蹄形湾。

(Previous Page) First Bend of the Jinsha River — Diqing, Yunnan
The Yangtze River makes its way to the terrain of abrupt changes in the alpine valley regions of the Yunnan-Guizhou Plateau after rushing down from the first step of the terrain. The river winds its way alongside the mountain. It is the first U-bend of the Yangtze River (the upper stream of which known as the Jinsha River) upon entering this region.

（从左至右）
怒江峡谷 · 云南福贡
滇池浪 · 云南昆明
喀斯特湿地 · 贵州荔波
湿地里的朱鹮 · 陕西汉中
玉龙雪山下 · 云南丽江
(From Left to Right)
Valley of Nu River — Fugong, Yunnan
Tides in Dian Lake — Kunming, Yunnan
The Karst Wetland — Shennongjia, Hubei
Crested Ibis in Wetlands — Hanzhong, Shaanxi
Foot of Mount Yulong — Lijiang, Yunnan

Yunan-Guizhou Plateau Wetland Zone

Yunnan-Guizhou Plateau Wetland Zone is situated in the southeast China. With the Qinling Mountains as its northern boundary, it boarders the Qinghai-Tibet Plateau to the west, and the Wu Mountains and the Xuefeng Mountains to the east. Stretching south to the northern edge of China's tropics, it covers the Yunnan-Guizhou Plateau, the Sichuan Basin and the part of southeastern Tibet. The whole area belongs to the second step of China's terrain, resting within the subtropical climate zone. Despite its subtropical monsoon climate, this region enjoys a mild whether, where the temperature and humidity fluctuates slightly from summer to winter due to its low latitude and high altitude. Great rivers starting from the first step, such as the Yangtze River, the Lantsang River and the Nu River, flow into the aforesaid region. Owing to the vast amount of precipitation brought by the southwest monsoon and the abrupt change in terrain, there are also many large or international rivers rising in this region, for instance, the Pearl River as the third largest river in China, the Han River and the Wu River as tributaries of the Yangtze River, and the Yuan River as the upper course of the Red River. Intense tectonic movements in the Yunnan-Guizhou Plateau leave a cross distribution of alpine and gorge regions, tiny flat dams and basins. A number of lakes often appear in fault zones or the watershed regions of major drainage systems. Many tributaries flow into the lakes while few water systems flow out. Thus, the residence periods of these lakes are greatly prolonged, causing the ecosystem fragile. Although most of these lakes are of fresh water, their degrees of mineralization evidently overwhelm that of the lakes on the eastern plains. The southwest monsoon brings ample amount of rainfall in summer, constituting more than 80% of the annual precipitation. The water level of lakes experiences large seasonal variation, with no ice conditions in winter. There are both mesotrophic and oligotrophic lakes in this zone. Many endemic species, such as *Viviparidae, Ottelia, Anoectochilus* and *Schizothorax*, dwell in this region. Because of the mild weather, developed swamps here mainly consist of reed swamps and carex marshes. Sphagnum bogs are rarely seen in this region, except in places such as Beihai of Tengchong County. Dashanbao in northeast Yunnan Province, Napahai in northwest Yunnan Province and Caohai in northwest Guizhou Province are important wintering grounds for black-necked cranes. The Dian Lake area is the major wintering ground for black-headed gulls.

There are many water systems flowing through the region. As the link between higher and lower regions, the landscape here stores immense water power resources. Mega projects like the South-North Water Transfer Project and the West-East Electricity Transmission Project are playing an important role in the transformation of wetlands in this region, but environmental impact assessment often lags behind and is impotent. Local wetland transformation will exert long-term and significant impact. The projects will greatly influence the energy source, water distribution as well as ecological patterns in China. Therefore, sound layout, environmental assessment and strategic planning at the state level are of particular importance. Laches and neglect of any kind cannot be allowed.

怒江第一湾 · 云南贡山

这是怒江进入云贵高原之后的第一大拐弯。怒江与金沙江、澜沧江三条大江从青藏高原奔腾而下，在横断山脉纵谷地区并流几百千米而不交汇，形成世界上罕见的“三江并流”的自然奇观，已列为世界自然遗产地。一条生态河流存在的意义，绝不仅仅是河水的奔涌流量，它养育着整个流域——高山、峡谷、沼泽、草甸、森林、湖泊及其万千生物。

First Bend of Nu River — Gongshan, Yunnan

It is the first large turn of the Nu River after it entering into the Yunnan-Guizhou Plateau. The Nu River, together with the Jinsha River and the Lantsang River, roars down from the Qinghai-Tibet Plateau. The three rivers run parallel in longitudinal valleys of the Hengduan Mountains for hundreds of kilometers without confluence, constituting an unusual landscape in the world, and listed as a World Natural Heritage Site. The existence of an ecological river is not only defined by the flushing water, but also by its nourishing of the whole drainage basin: mountains, valleys, swamps, meadows, forests, lakes and a myriad of lives.

大写的家“乡”· 云南丽江

峡谷中的澜沧江。

River Bend in Shape of the Chinese Character “hometown” — Lijiang, Yunnan

The Lantsang River in the valley.

秦岭南坡汉江源・陕西佛坪

呈东西走向的秦岭及淮河是中国地理上最重要的南北分界线。秦岭是长江流域与黄河流域的分水岭，是我国亚热带和温带的重要分界线，也是世界动物区系古北界和东洋界的分界线。秦岭南北坡的自然景观差异明显，南边过来的暖湿气流在这里受到秦岭的阻挡，降水充沛，山谷流水潺潺。这里是长江最大支流汉江的主要水源区，也是南水北调中线工程的主要汇水区域。

Head of the Han River at the Northern Slope of the Qinling Mountains — Foping, Shaanxi

Running east-west, the Qinling Mountains and the Huai River draw the most important dividing line of north and south in China's geography. The Qinling Mountains form the very watershed between the Yangtze River Basin and the Yellow River Basin, and serve not only as a significant boundary of subtropics and the temperate zone in China, but also as a dividing line between Palearctic and Oriental Region in respect of fauna distribution. The natural landscape varies a lot between that of the northern and southern slopes of the Qinling Mountains. Warm and humid air coming from the south is blocked here by the Qinling Mountains, bringing plentiful precipitation and developing gurgling streams across valleys. This region is the main water source of the Han River, namely the largest tributary of the Yangtze River. It is also a major water catchment region of the middle line of the South North Water Transfer Project.

金沙江虎跳峡・云南香格里拉

金沙江此处的两岸山岭和江面相差 2500 ~ 3000 米，落差巨大，谷坡陡峭，蔚为壮观，水势汹涌，声闻数里，为世界上最深的大峡谷之一。该处江心有一个十多米高的大石叫虎跳石，巨石犹如孤峰突起，屹然独尊，江流与巨石相互搏击，山轰谷鸣，气势非凡。

Tiger Leaping Gorge on Jinsha River — Shangri-La, Yunnan

Cliffs by the water rise 2,500 to 3,000 meters above the surface of the Jinsha River in this place. It is one of the deepest valleys in the world with the spectacular view of sheer valleys going all the way down at immense depth, and the booming sound of gushing water that can be heard miles away. A giant stone over ten meters high with the name "Tiger Leaping Stone" stands in the middle of the river. It rises like a lonely peak, overlooking everything else around. The magnificent thundering of the river, colliding with the giant stone, echoes in mountains and valleys.

川西的沼泽河湖·四川阿坝

在一定的自然环境条件下，河生湖、湖生沼泽，这里沼泽河湖在一个小范围里同时存在，共生共荣。

Swamps, Rivers and Lakes in West Sichuan Province — Aba (Ngawa), Sichuan

Under certain natural and environmental conditions, rivers can turn into lakes, and lakes, swamps. In this particular region, swamps, rivers and lakes coexist and reinforce each other.

滇西北高山湿地·云南高黎贡山西坡

高黎贡山属我国横断山的最西部，巨大的南北向山体连结了东西向的喜马拉雅山脉，阻挡了西北寒流的侵袭，又留住了印度洋的暖湿气流，形成了典型的亚热带气候。高黎贡山有着极典型的高山峡谷自然地理垂直带景观和丰富多样的动植物资源。山间降雨充沛，山谷平缓处的积水形成了美丽的高山湿地，并有沼泽产生。

Alpine Wetlands in Northwest Yunnan Province — West Slope of Gaoligong Mountain, Yunnan

The Gaoligong Mountain lies in the west end of the Henduan Mountains in China. Its giant north-south ranges connecting with the Himalaya Mountains block the incursion of northwest cold wave while retaining the warm and humid current from the Indian Ocean, forming a typical subtropical climate. A classic landscape of natural and geological altitudinal belts of alpine and gorge region as well as abundant fauna and flora resources bestow the Gaoligong Mountain. Plentiful precipitation in the mountains stops in gentle regions of valleys, forming alpine wetlands of great beauty and swamps.

高峡出平湖 · 湖北秭归

西陵峡是长江三峡中最险之峡，以滩多水急著称，三峡大坝蓄水后成为平湖。三峡水利工程在取得防洪、发电、交通等巨大经济效益的同时，也极大地影响了该地区以至长江的生态状况。图上消落带上限为海拔 175 米的三峡库区最高蓄水高程线。

Lake of Peace between high Gorges — Zigui, Hubei

The Xiling Gorge, famous for its numerous shoal and rapid was the direst gorge among the Three Gorges. After the impoundment of the Three Gorges Dam, the Xiling Gorge became a lake of peace. As huge economic benefits of flood control, electricity generation and communication being gained, the dam also greatly affects the ecological regime of local areas up to the Yangtze River. The upper limit of the hydro-fluctuation belt in the picture is the highest water impounding elevation at an altitude of 175 meters in the area of the Three Gorge Dam.

五花湖 · 四川九寨沟
清澈的湖水由于富含矿物质和水下生长有不同的植物，在山坡秋色森林的映衬下显得格外美丽。

Wuhua Lake — Jiuzhaigou, Sichuan
With rich minerals in the water and various plants living under it, the clear lake water shines beautiful against forests on the hillside in autumn.

欢腾的山涧 · 贵州黔西南
在水光热条件都非常丰富的亚热带地区，山涧流水在葱绿繁茂多样的植物丛中跌宕起伏，呈现出的画面极具生命力及自然美感。

Mountain Streams of Rejoice — Qianxinan, Guizhou
In subtropical areas, where water, light and thermal conditions are sound, a mountain stream winds its way across lush plants of various kinds. It is a view of enormous vitality and natural beauty.

岩溶湿地之大成 · 云南丘北普者黑

中国的岩溶（也称喀斯特）地貌分布之广、类型之多为世界罕见。在中国，裸露的碳酸盐类岩石面积约 130 万平方千米，约占全国总面积的 1/7，云贵高原及其边缘地带是世界上最大的岩溶地貌分布区之一。裸露的碳酸盐类岩石在水的作用下形成的地貌，不仅是地表破碎，形成峰林、峰丛、漏斗等。地下也破碎复杂，形成裂隙、溶沟、溶洞以至地下水系等。由于岩溶“漏水”的地质结构特性，保护岩溶湖泊及其物种资源、生态环境就尤显重要。云南普者黑的岩溶景观颇具代表性，它是一个由幽静秀丽的高原湖泊群，苍翠叠嶂的峰丛峰林，鬼斧神工的各种溶洞和峡谷，丰富多样的动植物种以及绚丽多彩的民族民俗文化巧妙融为一体的岩溶湿地生态系统大成。

Masterpiece of Karst Wetlands — Puzhehei of Qiubei, Yunnan

Carbonate rocks has reached nearly 1.3 million km^2, accounting for one seventh of the total land area of China. The Yunnan-Guizhou Pleateau and the bordering area is one of the largest regions distributed with Karst landscape. The landscape formed by bare Carbonate rocks under the action of water is not only fragmented on the surface, featuring "stone forest", "stone clusters" and funnels, but also complicated underground with numerous cracks, gullies, caves and under water systems. Due to the "water leaking" characteristics of the Karst landscape, it is extremely important to protect Karst lakes and their species resources and ecological environment. The Karst landscape in Puzhehei of Yunnan province is very representative. It is a masterpiece of Karst wetland ecosystem, a perfect combination of peaceful and beautiful plateau lake groups, towering and green "stone forest" and "stone clusters", uncanny caves and gorges, diverse species of wild fauna and flora and the magnificent ethnic folk arts and cultures.

1

2

3

4

1. 漂浮状苔草沼泽 · 云南腾冲北海

一米多厚的草垫像一片片浮毯漂浮在水面上，当地人常把随意切开的一小块草排当竹筏来划，抓捕鱼虾。

Floating Carex Marshes — Beihai, Tengchong, Yunnan

Grass mats over a meter thick float on the water surface like many pieces of floating blankets. Local residents often cut a patch of it and ride on it as a fishing raft.

2. 喀斯特森林中激流 · 贵州茂兰

在喀斯特地区，水流或在地下，或溢出地表，形成了岩溶泉、断头河、水中森林等特殊的湿地形式。

Torrents in Karst Forests — Maolan, Guizhou

In karst regions, water flows under or above ground, forming special forms of wetlands, such as karst springs, beheaded rivers and forest in water.

3. 多彩的钙化池 · 四川黄龙

由于钙化池的特殊环境条件，池水的洁净度和透明度极高。

Multicolored Calcified Ponds — Huanglong, Sichuan

The unique environmental conditions of calcified ponds give rise to water of extremely high cleanliness and transparency.

4. 地热湿地 · 云南腾冲

在高温水的环境里，同样有绿色的生命存在。

Geothermal Wetlands — Tengchong, Yunnan

Green life can exist even in high-temperature water.

冷杉森林沼泽 · 四川王朗

冷杉林地中在土壤过度潮湿甚至积水的情况下形成的沼泽。沼泽这种类型在本区不多见，但却是本区域很有特色的湿地景观。

Fir Forest Swamps — Wanglang, Sichuan

Fir forests form such swamps when the soil becomes too wet or even water-logged. Swamps are not commonly spotted in the region, but they are a kind of distinctive wetland landscape among local areas.

水清鱼稠 · 贵州黔南

由于植被茂密，西南山地的山涧水溪水质尤为清澈，云南光唇鱼自由自在地生活在这里，这种美丽的湿地景观，颠覆了“水清则无鱼”的俗传。

Clear Water Rich with Fish — Qiannan, Guizhou

The creeks of southwest mountains are particularly limpid, thanks to the dense vegetation. The *acrossocheilus yunnanensis* live blithely in the water. The fair wetland scenery overturns a Chinese folklore: pure water holds no fish.

湿地的灵动 · 重庆长寿

生于浅水中的杉叶藻属多年生的挺水或沉水植物，无分枝的茎挺出水面摇曳，像无数少女在舞台上翩翩起舞，使湿地充满了美感和灵动。

Dynamic Wetlands — Changshou, Chongqing

Growing in shallow water, the *hippuris* are perennial emerged plants or submerged plants. Their unbranched stocks sway out of the water, like numerous maidens dancing on the stage, endowing the wetlands with beauty and dynamism.

岩溶洞穴地下河 · 广西百色

岩溶洞穴水系是湿地的一种特殊类型，它是具有一定汇水范围的、由主流及各级支流构成的岩溶地下水流。本地区由于遍布岩溶地貌，加之降水充沛，岩溶洞穴水系相当发达。岩溶洞穴水系是河流水系（如珠江水系）不可分割的一部分。

Subterranean Streams in Karst Caves — Baise (Baksaek), Guangxi

The subterranean water system in karst caves is a special type of wetland. It is a subterranean river consists of a main stream and tributaries of different levels with certain catchment areas. Karst landform and ample precipitation in this area give rise to a well-developed karst cave drainage system. The system is an integral part of river systems, such as the Pearl River system.

情人潭的水真的很迷人 · 湖北十堰

在中国地势的二级阶梯向三级阶梯过渡地带，由于地形及环境原因，往往热量、雨量都十分丰富，万物生长、植被茂密。在山林河谷、沟谷中产生了很多山涧、溪流，同时也伴有不少瀑布、深潭，水声哗哗、水质清澈。图为情人潭，透过密林的阳光直射入深潭，潭水虽清但深不见底，景色光怪陆离，十分梦幻迷人。

Charming Water in Lovers' Pond — Shiyan, Hubei

In the buffering region between the second to the third step in China's terrain, special terrestrial and environmental conditions give rise to rich heat and precipitation. Various forms of life thrive. Dense vegetation covers the land. Many creeks and streams, as well as a number of waterfalls and deep ponds, form in river valleys and ravines, in which clear water keeps rushing. The picture presents the Lovers' Pond. The sunbeam shines directly into the deep pond through the dense forest. Clear as the water is, the bottom of the pond still cannot be seen. The grotesque view has its charms and wonders.

我国的物种新记录——钳嘴鹳 · 云南蒙自

钳嘴鹳于 2006 年 10 月在云南洱源西湖首次记录，之后陆续在云贵高原多处发现。钳嘴鹳来自东南亚，数量有越来越多的趋势，有些甚至成了留鸟。有专家推测，钳嘴鹳分布范围北扩与全球气候变暖带来的气候带北移有关。值得我们尤要注意的是，随着全球气候变暖带来的气候带北移趋势，今后从我国南面国家自然迁移过境来的物种还会越来越多，这种趋势对当地湿地生态系统以至整个生态系统产生的影响不可忽视，亟待科学的评估和对策。

Newly Recorded Species in China — Asian Open-bill Stork, Mengzi, Yunnan

The Asian open-bill stork was first recorded in Eryuan West Lake, Yunnan Province in October 2006. Afterwards, it was spotted in many places on the Yunnan-Guizhou Plateau. The storks come from Southeast Asia, and on increasing trends, some of them become resident birds. Experts reckon that the northward distribution of Asian open-bill storks has something to do with the north shift of the climate zone due to global warming. We must pay special attention on the fact that with the climate zone's northward movement caused by global warming, a growing number of species travelling from countries in the south will arrive in China. Such tendency will impose significant influence on wetland ecosystems, and even on the whole ecological system. Scientific assessments and countermeasures are urgently needed.

1

2

3

4

5

6

1. 红嘴鸥 · 云南滇池
   Black-headed Gulls — Dian Lake, Yuannan

4. 水雉 · 云南大理
   Pheasant-tailed Jacana — Dali, Yunnan

2. 大鲵 · 重庆万州
   Giant Salamander — Wanzhou, Chongqing

5. 朱鹮 · 陕西洋县
   Crested Ibis — Yang County, Shaanxi

3. 凤头麦鸡 · 四川乐山
   Northern Lapwing — Leshan, Sichuan

6. 小䴙䴘 · 四川宜宾
   Little Grebes — Yibin, Sichuan

高原黑颈鹤 · 云南大山包
从若尔盖飞来越冬的黑颈鹤在云贵高原的阳光中翱翔。

Blacked-necked Cranes on Plateaus — Dashanbao, Yunnan
Blacked-necked cranes, migrating from Zoigê County to spend the winter, soar in the sunshine on the Yunnan-Guizhou Plateau.

春田水暖 · 云南元阳

哀牢山梯田是哈尼人民千百年来社会可持续发展的美丽家园，哈尼人开垦的梯田随山势地形变化而因地制宜，大者数亩，小者仅有簸箕大，一坡梯田的最高级数达 3000 多级。哈尼梯田是森林 — 水溪 — 村寨 — 梯田 — 鱼鸭五度同构的，是一个人与自然高度协调、良性循环的人工湿地生态系统。

Warm Water on Terrace Fields in Spring — Yuanyang, Yunnan

The terrace fields in the Ailao Mountains are the beautiful homeland to the Hani people for thousands of years, thanks to their sustainable social development. The Hani people adjust soundly to the topographic conditions of local mountains in the reclamation of terrace fields, with large ones covering several acres and small ones only the size of a dustpan. A slope of terrace field can be of over 3,000 steps. The Hani terrace fields combine five elements together, namely forests, streams, villages, terrace fields as well as fishes and ducks. It is a constructed wetland system of virtuous circle, well-balanced between man and nature.

鸥满春城 · 云南昆明

红嘴鸥已经成为了与春城人民和睦相处的常客，每年有 1 万多只从西伯利亚来到这里过冬，这种人鸟和谐相处的景象，在全国大中城市中是十分难得的。

Spring City Filled with Gulls — Kunming, Yunnan

Black-headed gulls have become frequent visitors in Kunming. They get along with residents of the Spring City. More than ten thousand of them, migrating from Siberia, choose here as their annual wintering ground. The harmonious coexistence man and birds can hardly be seen in large-and-medium cities across the country.

黑颈鹤姑娘 · 云南大山包

多少年来，自然保护区保护站的姑娘与越冬来的黑颈鹤朝夕相处，保护它、救护它，贡献了宝贵的青春。

Black-necked Crane Girl — Dashanbao, Yunnan

For years, girls from natural reserve stations live with the wintering black-necked cranes from dawn till dusk. They dedicate their precious youth to protect and rescue these endangered birds.

加高了大坝的丹江口水库 · 湖北十堰

这里是中国南水北调中线工程的水源地。修建在汉江上的丹江口水库大坝加高后（图上大坝上部的直立部分），正常蓄水位将从 157 米提高至 170 米，库容从 174.5 亿立方米增加到 290.5 亿立方米，库区成了亚洲第一大人工淡水湖。中线一期工程年均可向河南、河北、北京、天津等四省（直辖市）的 20 多座城市调水 95 亿立方米，中远期规划每年调水量将达 130 亿立方米，将有效缓解中国北方水资源严重短缺的局面。

Danjiangkou Reservoir with Dam Heightened — Shiyan, Hubei

This is the water source of the middle Route of the South-North Water Transfer Project. The dam of Danjiangkou Reservoir on the Han River has been heightened (the vertical upper part of the dam in the picture), allowing normal pool level to rise from 157 to 170 meters, and the reservoir capacity to enhance from 17.45 to 29.05 billion cubic meters, making the reservoir area the largest artificial freshwater lake in Asia. The first stage of the middle route project can transfer 9.5 billion cubic meters of fresh water to more than 20 cities in Henan, Hebei, Beijing and Tianjin. The mid- and long-term project will enlarge water transfer to 13 billion cubic meters, which will effectively relieve the severe water shortage in northern China.

澜沧江漫湾电站 · 云南景东

西南地区由于上承青藏高原、下接平原丘陵，河流坡降很大，加之低纬度地区的充沛降水，水利资源异常丰富，因而这个地区已建和欲建的大型水利工程星罗棋布，一条江上有多达二十几级的梯级电站和拦河大坝的建设规划。这些水利工程很大程度上影响着中国的能源、水分布及生态格局。因此，国家层面上的科学布局、环境影响评价、战略规划尤为重要。

Manwan Power Plant on the Lantsang River — Jingdong, Yunnan

The southwest China connects the Qinghai-Tibet Plateau to hilly plains. Large river gradients in the region, together with ample precipitation of low latitudes, bring extremely abundant hydro-power resources. Thus, large-scale water projects built or to be built scatter in this region. There are construction plans of more than 20 steps of cascade power stations and dams on a single river. These water projects will greatly influence the energy source, water distribution as well as ecological patterns in China. Therefore, sound layout, environmental impact assessment and strategic planning at the state level are of particular importance.

4 东北、华北平原湿地区

Northeast and North China Plains Wetland Zone

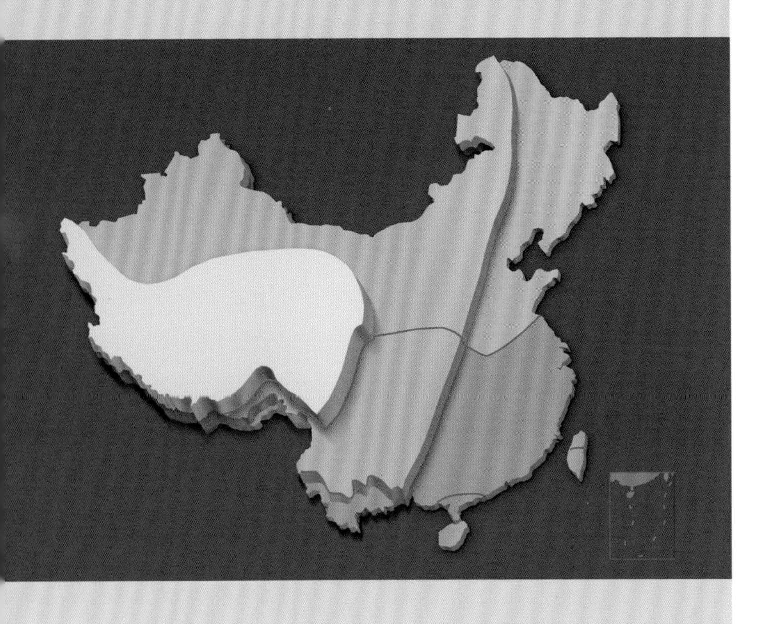

(前页）湿地迷宫·黑龙江嫩江
这里看到的是河道、湖泊、沼泽、草地相互交织在一起的画面，要想一一判断分清还真不容易。其实，湿地就是这样，是分不清、理还乱的复合生态系统。

(Previous Page) A Maze of Wetlands — Nenjiang, Heilongjiang
An interweaving of river, lake, swamp and grassland is very difficult to distinguish. In fact wetland is such a complex ecosystem.

东北、华北平原湿地区

东北、华北平原湿地区为我国大兴安岭及太行山脉以东，淮河以北的地域。包括三江平原、松嫩平原、小兴安岭、长白山、辽河平原和华北平原范围及其近海和海岸、沿海诸岛，属中国地势第三级阶梯的北部分。气候区为温带，属温带季风气候。这里夏季高温多雨，冬季寒冷干燥，四季分明，冬夏季风方向变化显著。黑龙江、松花江、乌苏里江、嫩江、辽河及滦河、海河为本地区的主要河流，黄河从二级阶梯的黄土高原下来，经华北平原到渤海湾入海，还有人工湿地的京杭大运河(北段)。该地区特点是河流密布、有大片的湖沼湿地分布，成因多与地壳沉陷、地势低洼、排水不畅和河流摆动等因素有关，湖泊具有面积小、湖水浅、湖盆坡降平缓、现代沉积物深厚、矿化度较高等特点。分布于山区的湖泊成因多与火山活动有关，如长白山天池为火山口湖，镜泊湖和五大连池是典型的熔岩堰塞湖等，也有不少作为人工湿地的大中型水库。该区北段寒冷湿润的环境下，在大、小兴安岭和长白山地区分布有很多的森林沼泽，在三江平原产生了中国最大的沼泽河湖湿地，其中很大部分已经改造成了水稻田。从鸭绿江口到苏北沿海多为淤泥质、砂质海滩，滩涂平缓广阔，各种湿地植物繁茂，草丛中有丰富的有机质基底，为迁徙鸟类提供了理想的食物和隐蔽条件，是世界三大迁徙路线之一的东亚－澳大利亚鸟类迁徙路线上最重要的停歇地和栖息地。其水禽种类丰富，非常合适丹顶鹤、白头鹤、白枕鹤、黑鹳、中华秋沙鸭等鸥类、雁鸭类、鸻鹬类水禽的栖息繁殖。辽东湾的双台子河口地区是黑嘴鸥的重要繁殖地，山东半岛的荣成海滨是我国北方沿海大天鹅的主要越冬地，新发现天津大港为遗鸥的主要越冬地之一。还有水獭、水貂和一些啮齿类动物等也生活在该区的沼泽草地生态系统中。

这里地势平缓，湿地星罗棋布，人们逐水而居，湿地水鸟种类繁多，是世界候鸟迁徙途径的最重要的区域之一，无论是创造中华文明繁荣的中原大地还是近百年内才大量开发崛起的东北平原，人与湿地在过去和现在都在谱写人类社会如何可持续发展的重大关系。

(从左至右）
河流、牛轭湖和农田综合湿地景观·内蒙古加格达奇
平原湖泊·黑龙江嫩江
三江平原塔头湿地·黑龙江抚远
水稻田里的牛背鹭·吉林延边
港湾冬天的早晨·河北秦皇岛
(From Left to Right)
A Wetland Landscape Integrating River, Oxbow Lake and Farmland — Jiagedaqi, Inner Mongolia
Lakes on the Plain — Nenjiang, Heilongjiang
Undergrowth Grassy Marshland on Sanjiang Plain — Fuyuan, Heilongjiang
Egrets Appear in Paddy Fields — Yanbian, Jilin
A Winter Morning in Bay — Qinhuangdao, Hebei

Northeast and North China Plains Wetland Zone

Northeast and North China Plains Wetland Zone is located to the east of Greater Xing'an Range, Taihang Mountains and north of Huai River, including Sanjiang Plain, Songnen Plain, Lesser Xing'an Range, Changbai Mountain, Liaohe Plain, North China Plain and subsequent coastal waters, coastlines and offshore islands which belong to the north part of the third step of China's terrain. Dominated by temperate monsoon climate, it is characterized by four distinctive seasons which include hot and rainy summer, cold and dry winter, and drastic change in wind directions between winter and summer seasons. Main rivers in this area include Heilong River, Songhua River, Ussuri River, Nen River, Liao River, Luan River, Hai River, and north part of the Grand Canal, a constructed wetland. In addition, the Yellow River flows down from the Loess Plateau at the second step and runs throughout the North China Plain into the Bohai Bay. This area is featured with dense rivers, large patches of lakes, swamps and marshes, mostly caused by descending crust, lower terrain, impeded drainage and swinging rivers. Lakes are usually in small area and highly mineralized, with shallow water, gentle slopes and piles of recent sediments in basin. Lakes in highlands are often caused by volcano activities, e.g. Tianchi in Changbai Mountain is a crater lake, Jingbohu and Wudalianchi are typical barrier lakes. Quite a number of large and medium-sized reservoirs are constructed wetlands. Dominated by the cold and humid climate in the north, many swamp forests are found in Greater and Lesser Xing'an Range and Changbai Mountain. China's biggest marsh wetland occurs in Sanjiang Plain, though a large portion has now been converted into paddy fields. Stretching from the mouth of Yalu River down to north of Jiangsu, the coastal zone of this area is mostly muddy, sandy beaches with gentle and wide tidal flats growing lush wetland plants of various types and grasslands rich in organic bases, thus providing ample food and ideal shelter for migratory birds. Being the most important stopover site and habitat in the East Asia and Australasia flyway, one of the world's three major flyways, it has plentiful waterfowl species, ideal for such gulls, ducks and waders as red-crowned crane, hooded crane, white-naped crane, black stork and Chinese merganser to inhabit and breed. The estuary at Shuangtaizi in Liaodong Bay is a critical breeding site for Saunder's Gull, the coasts along Rongcheng in Shandong Peninsula is a major wintering site for Whooper Swan in north China, and Dagang of Tianjin is a newly discovered wintering site for Relict Gull. Other animals such as otter, mink and rodents also inhabit in the marshes and grassland in this area.

As one of the major flyways in the world, this area, with numerous wetlands and various waterfowls, is flat in topography and people choose to inhabit by the water. Either in Central China which is the cradle of Chinese civilization, or in Northeast China Plain which has less than 100 year history of mass exploitation, man and wetland together are constantly depicting the concept of sustainable social development.

夕照天池 · 吉林长白山

长白山是松花江、图们江、鸭绿江的三江之源。长白山天池位于主峰火山锥体的顶部，是我国最大的火山口湖。天池水深达 370 余米，是中国最深的湖泊。从天池倾泻而下的长白飞瀑，是世界上落差最大的火山湖瀑布，天池周边还有多处温泉湿地。

Sunset over the Tianchi Lake — Changbai Moutain, Jilin

Changbai mountain is the source of Songhua River, Tumen River, Yalu River. Tianchi Lake, the largest crater lake in China, is located at the top of the main volcanic cone of Changbai Mountain with a water depth of 370 meters, the deepest one in China. Pouring down from Tianchi, Changbai waterfall is a volcanic lake waterfall with the biggest drop in the world. There are a number of hot springs wetlands around Tianchi.

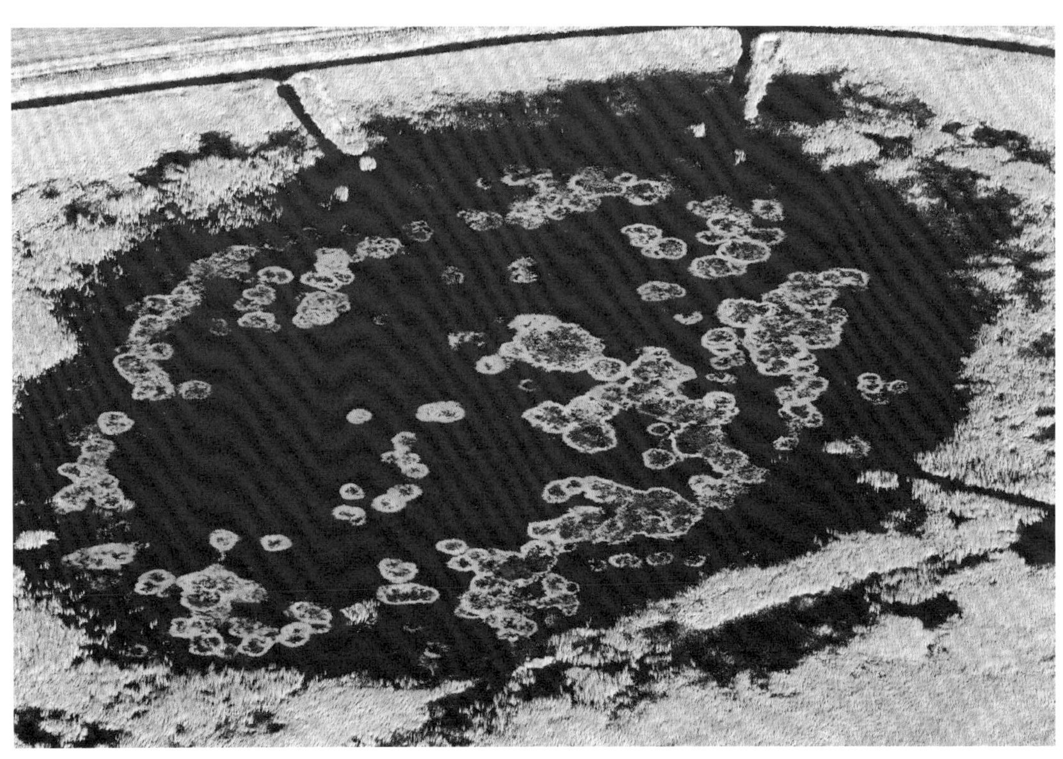

逐渐沼泽化的湖泊 · 黑龙江齐齐哈尔

自然界中的湖泊一方面经过长期的泥沙淤积、化学沉积等，湖水会逐渐变浅；另一方面在光照、温度等条件适宜的情况下，生长的喜光植物和漂浮植物会越来越多。由于死亡植物不断堆积湖底，分解很慢，植物残体逐年累积而形成泥炭。随着泥炭的增厚，湖水进一步变浅，湖面缩小，最后泥炭堆满湖盆，水面消失，水草丛生，就演化为沼泽，这是湖泊湿地自然演替的必然过程。

Bogging Lakes — Qiqihar, Heilongjiang

Lakes will become shallow due to siltation and chemical deposition over time. Meanwhile more and more heliophile and floating plants grow under favorable sunlight and temperature in the lake. Dead plant slow in decomposition will settle at the bottom of the lake and turn into peat moss after many years of accumulation. When lake becomes shallower and the lake surface shrinks due to increased peat mosses, heaps of peat moss will cover the lake basin. Eventually water disappears, water plant overgrows and marsh appears. This is the natural succession of lake wetlands.

寒温带湿地大成 · 黑龙江南瓮河

在南瓮河湿地建立了我国最重要的以寒温带湿地生态系统为保护对象的自然保护区。区内的原始森林湿地、草丛湿地、灌丛湿地、冰湖湿地、岛状林湿地、牛轭湖等河流、湖泊、沼泽湿地齐全完整，河道尤其蜿蜒迂回曲折，景观极其美丽迷人复杂，几乎囊括有我国寒温带陆生、湿生、水生生物物种，其完整、独特的生态系统在国内极为罕见。

A major wetland in cold temperate zone — Nanweng River, Heilongjiang

Nanweng river wetland is the most important nature reserve to protect the wetland ecosystems in the cold temperate zone. It is home to a wide variety of different types of rivers, lakes and marshes such as the primitive forest wetlands, sedge marshes, shrub swamps, ice lake wetland, and isolated wood wetlands. The winding and meandering rivers present a marvelous and captivating landscape with a wide range of the species in the cold temperate zone, from terrestrial species, wetlands species to aquatic organisms. Its complete and unique ecosystems are extremely rare in the country.

“小老头树”沼泽 · 吉林白山

在东北有冻土层分布的地段，因水分下渗困难，地表过湿形成了沼泽。因恶劣的生长环境限制了树木的正常发育，或造成树木大量死亡而形成“站杆”；或出现树木枯梢、生长缓慢的“小老头树”。这种现象，在大、小兴安岭和长白山林区都可以看到。

Swamps with Small Aged Trees — Baishan, Jilin

In Northeast China where there are frozen soil layer, saturated land surface becomes swamp due to impeded water infiltration. This harsh environment restricts the normal growth of trees, thus results in either dead trees still standing like poles, or small aged trees suffering from die-back and growing very slow. This can be seen in Greater and Lesser Khingan Range and Changbai Mountains forest regions.

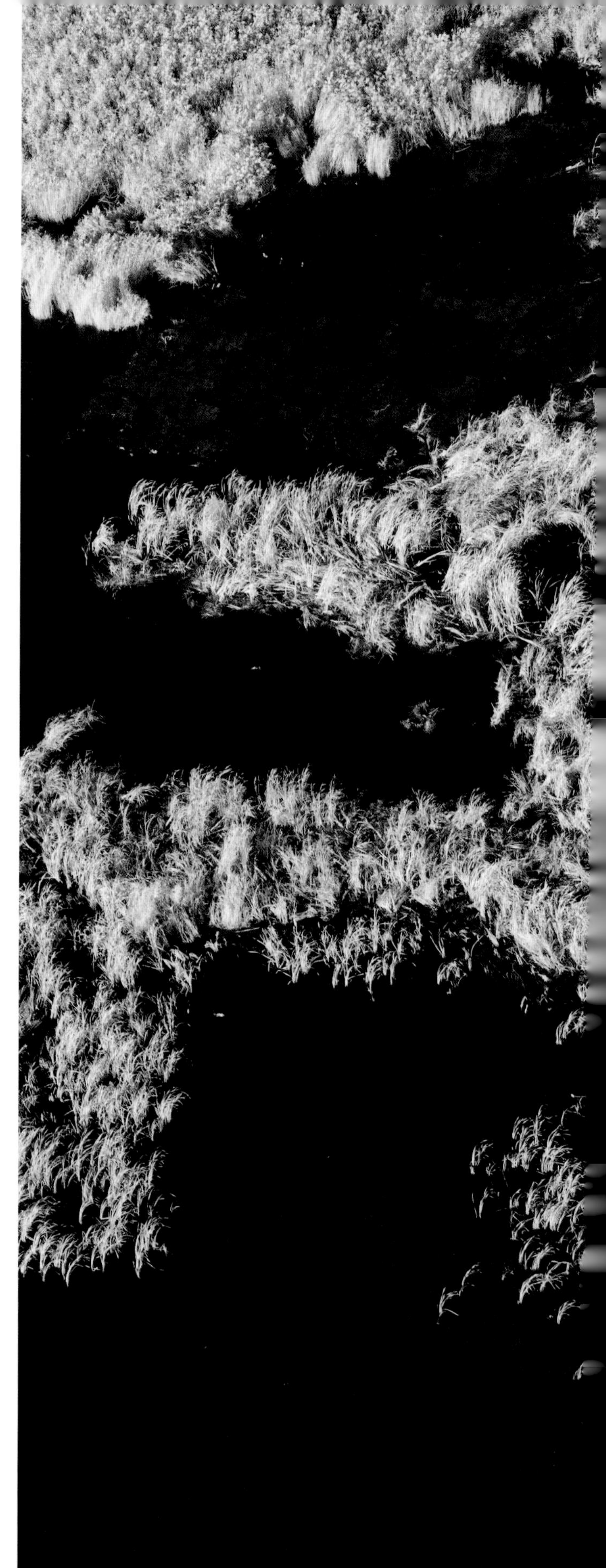

苔草沼泽与水鸟 · 黑龙江黑河

东北平原地区是我国沼泽主要分布区之一，沼泽面积约 300 万公顷，占东北地区总面积的 2.7% 强，占全国沼泽面积的 1/3 左右。这些地表多年积水或土壤处于过湿状态，主要生长有苔草、芦苇、小叶章等沼生和湿生植物，并有泥炭的形成和累积，土壤具有明显的潜育层。这里是水鸟上佳的停歇地和觅食地。

Sedge Marsh and Waterfowls — Heihe, Heilongjiang

Northeast China Plain is the major region in China where marshes distribute, with an area of around 3 million hectares, accounting for one third of the total marshes in China and 2.7% of the land area in northeast China. Aquatic plants such as sedge, reed and narrowleaf small reed grow on the hydric soil and land surface saturated all year round here. Soil has apparent gley horizon nature with peat forms and accumulates. This is the best place for waterfowls to stop over and forage.

珍宝岛湿地 · 黑龙江虎林

乌苏里江在河滩上留下不少故河道，逶迤迂回，泡沼密布，而地表植被多以小叶章为主。有的故河道沙坝被茂密的蒙古栎等乔木覆盖，构成了奇特无比的岛状林景观。这里是三江平原保存最完好的原始湿地之一。

Wetlands on Zhenbao Island — Hulin, Heilongjiang

Zigzagging old courses left by Ussuri River on the river bank are covered by bogs and ground plants, mainly narrowleaf small reed. Sandy banks of the old courses covered by dense Mongolian oak forest and other species form a unique island forest landscape. This is the best preserved pristine wetland among those in the Sanjiang Plain.

红绿海滩 · 辽宁盘锦

在辽河入海口，由于淡水携带大量营养物质沉积并在潮水反复浸淹作用下，形成了适宜多种生物繁衍的河口湾湿地。河口湾大面积的滩涂上，生长着翅碱蓬这种潮间带植物。每年 4 月份起，翅碱蓬由绿慢慢变红，到了 9 月份便像落下的红霞在海陆间燃烧。绿色的芦苇、红色的翅碱蓬交相映衬，显得格外奇特美丽。这里是世界濒危鸟类——黑嘴鸥最大的繁殖地。

Green and Reddish Coast — Panjin, Liaoning

Estuarine wetland favorable for various living organisms to inhabit occurs at the estuary of Liao River due to the sedimentation of nutrients brought by the freshwater and repeated inundation of tides. Seepweed, a tidal plant, grows on the large tidal flats along the estuary. From April each year, the green color gradually turns red, and in September it looks like burning rosy clouds. The different color of green reeds and red seepweed enriches this beautiful landscape. This is the largest breeding site of Saunder's Gull, a world-class endangered species.

滩涂的开发与保护 · 辽宁盘锦

包括鸭绿江、辽河、海河、黄河等河口在内的整个渤海湾滩涂，是东亚－澳大利亚候鸟迁徙的大通道，同时也是我国人口密集、工业密集、油田及港口开发的重点地区，开发与保护的矛盾十分突出。

Development and Conservation of Tidal Shoal — Panjin, Liaoning

The whole Bohai tidal flats including the estuaries of the Yalu river, Liao river, Haihe river and the Yellow River serve as the flyway for east Asian– Australian migratory birds, and is also densely populated for industry, oil field and port development at the same time, therefore, the contradiction between development and protection is very acute.

晚归丹顶鹤·黑龙江扎龙

丹顶鹤姿态优雅，体色分明，有“湿地之神”的美称，东亚地区的居民，用丹顶鹤象征幸福、吉祥、长寿和忠贞，在各国的文学和美术作品中屡有出现。

Red-crowned Cranes Going Home — Zhalong, Heilongjiang

Red-crowned Crane is reputed as Fairy of Wetland for its elegant appearance and distinguished body color. Being taken by people in East Asia as a symbol of luck, happiness, longevity and fidelity, this bird frequently appears in art and literature works in these countries.

白枕鹤振翅 · 内蒙古锡林郭勒

每年 4 月上、中旬，在白枕鹤求偶、筑巢的繁殖前期，鹤的领域中常发生鹤与鹤之间的对峙行为。这时已经占有领域的白枕鹤，会时刻警惕注视着入侵者，着急时就会双双大声鸣叫，雄鹤仰头振翅、雌鹤也仰头鸣叫，有“鹤鸣千里”一说。

White-naped crane flapping the wings — Xilingol, Inner Mongolia

During the middle of April every year, at the early stage of the white-naped crane courtship, nesting and breeding, there are standoffs going on often among the cranes. The white-naped cranes which have already occupied a territory will be vigilant about the invaders. If being offended, the couple will both sing loudly, the male flapping wings while female lifting their heads and screams, from which derives the saying of "the scream of cranes could be heard a thousand miles away".

飞过芦苇湿地的丹顶鹤·吉林向海

丹顶鹤的繁殖地在中国东北的松嫩平原、三江平原，俄罗斯的远东和日本等地，而越冬在中国的东南沿海及长江下游、朝鲜半岛和日本等地。

Red-crowned Cranes Flying over Reed Bed — Xianghai, Jilin

Red-crowned Crane breeds in Songnen Plain, Sanjiang Plain of Northeast China, Russia's Siberia and Japan, winters in lower reaches of Yangtze River and southeast coasts of China, Korea Peninsula and Japan.

滩涂上的鸻鹬类鸟 · 辽宁鸭绿江口

鸻鹬类是地球上迁徙距离最远的鸟类，我国的东海岸是全球三大鸟类迁徙路线之一的东亚—澳大利亚迁徙路线上的关键地段，迁徙的鸟类中以鸻鹬类鸟数量最大、种类最多。

Waders on the Shoal — Yalu River Estuary, Liaoning

Waders are birds on earth that could migrate the longest distance. As a critical part in the East Asia and Australasia Flyway, one of three major flyways in the world, China's east sea coastal zones are featured for the biggest number of wader species and population.

1

2

3

4

5

6

7

8

9

1. 泽鹬
   Marsh Sandpipers

2. 鹤鹬
   Spotted Redshank

3. 反嘴鹬
   Pied Avocet

4. 林鹬
   Wood Sandpiper

5. 斑尾塍鹬
   Bar-tailed Godwit

6. 金眶鸻
   Little Ringed Plover

7. 黑翅长脚鹬
   Black-winged Stilt

8. 白腰杓鹬
   Eurasian Curlews

9. 黑尾塍鹬
   Black-tailed Godwit

湿地中的一对狍子 · 黑龙江大兴安岭
不仅是鸟类，很多兽类的生活与湿地也是息息相关的。
A Pair of Roe Deer in the Swamp— Greater Khingan Range, Heilongjiang
Besides birds, many mammals also have closely relationship with wetlands.

1

2

3

4

5

6

7

1. 鸳鸯 · 河南三门峡
Mandarin Duck — Sanmenxia, Henan

2. 鸿雁 · 河北唐海
Swan Goose — Tanghai, Hebei

3. 水獭 · 黑龙江鸡西
Otter — Jixi, Heilongjiang

4. 须浮鸥 · 河北衡水湖
Whiskered Tern — Hengshui Lake, Hebei

5. 白枕鹤 · 天津北大港
White-naped Crane — Beidagang, Tianjin

6. 疣鼻天鹅 · 山东黄河三角洲
Mute Swan — Yellow River Delta, Shandong

7. 小天鹅 · 山东荣成
Tundra Swan — Rongcheng, Shandong

苍鹭・河南郑州

苍鹭，我国湿地中最为常见的一种大型涉禽，又称"老等"，因它们不论觅食还是休息，始终都保持不慌不忙甚至不动的状态而得名。

Heron — Zhengzhou, Henan

Heron, one of the most common large waders in the wetland, also called "always waiting", because no matter when they feed or rest, they always keep unhurried or sometimes even motionless state.

芦苇丛中的白琵鹭・山东黄河三角洲

特有的琵琶型的大嘴是琵鹭类鸟共有的特征，它更适于发掘滩涂浅层的食物。白琵鹭喜欢成群活动，这和喜欢独处的苍鹭不同。但是它们都有个共同点，在南方繁殖的种群不迁徙，为留鸟，在东北等寒冷地方繁殖的种群通常都要到南方越冬，为候鸟。

White spoonbill in reeds — Yellow River Delta, Shandong

The type of spoon-shape mouth is a common characteristic of spoonbill bird. It is more suitable for the excavation of food from shallow tidal flats. White spoonbills like swarms activities, which is different from herons that prefers solitude. But they all have one thing in common, those that reproduce in the south do not migrate, and are known as the resident birds; those that reproduce in the cold places such as the northeast generally winter in the south, and are known as migratory birds.

风雪天鹅港·山东荣成

这里是亚洲最大的大天鹅越冬栖息地。适宜的气候、充足的食物、干净的水源和优美的环境使荣成拥有了这份殊荣，但更重要的原因是这里的人们对野生动物的百般呵护。

Swan at the Snowy Port — Rongcheng, Shandong

This is the largest wintering site for whooper swan in Asia. Suitable climate, abundant food, clean water and nice surroundings grant Rongcheng this reward. But most importantly, it is because local people love and care about wildlife.

皇家园林落天鹅 · 北京颐和园

冬天来临了，成群的大天鹅降临颐和园，它们在这里停歇、觅食和补充体力，自由自在地生活在昔日皇家园林的昆明湖上，人们关注它们、小心地呵护它们，直至天鹅自由地离去。

Swan Stops in Royal Garden — Summer Place, Beijing

The winter comes and flocks of swan which stop in Summer Place to rest, forage and enhance physical strength, enjoying undisturbed life on the Kunming Lake in this former royal garden. They have been taken good care by the local residents before their departure.

京杭大运河北端——通州运河及古运粮楼 · 北京通州

京杭大运河是世界上开凿最早、最长的一条人工河道，在中华民族的发展史上，为沟通南北经济和水运做出了巨大的贡献。现今，南水北调东线工程利用京杭大运河以及与其平行的河道输水，连通洪泽湖、骆马湖、南四湖、东平湖作为调蓄水库。 为缓解中国北方严重缺水的状况，古老的大运河再次焕发了青春。

Tongzhou Canal and the Ancient Grain-transportation Tower in Northern Part of the Hangzhou-Beijing Grand Canal — Tongzhou, Beijing

Being the world's first and longest waterway constructed by man, the Hangzhou-Beijing Grand Canal has made tremendous contributions to the economic ties and water transportation between north and south in the Chinese history. Nowadays the Grand Canal is making renewed contribution to alleviate serious water shortage in the north by drawing water together with its parallel river courses, and connecting Hongze Lake, Luoma Lake, Nansi Lake and Dongping Lake as regulating reservoirs in the eastern section of the South-to-North Water Diversion Project.

5 华东、华中、华南平原丘陵湿地区

East, Central and South China Plains and Hilly Areas Wetland Zone

(前页）江南古村水情长 · 安徽宏村
宏村是世界文化遗产地，被誉为"中国画里的乡村"。在长江以南的苏南、浙北、皖南、赣东北地区，自古就享有人间天堂之美誉。这里河湖交错，水网纵横，小桥流水，田园村舍，如诗如画，融自然景观和人文景观为一体，是人与湿地协调发展的典范。

(Previous Page) Water Flowing in an Ancient Village — Hongcun, Anhui
A UNESCO World Heritage Site, Hongcun is noted as a village in a Chinese painting. Traditionally south Jiangsu, north Zhejiang, south Anhui and northeast Jiangxi at the south of the Yangtze River are reputed as Heaven on Earth. Interweaving rivers and lakes, stone bridges and village houses here forms a picturesque pastoral life that demonstrates the development of both man and wetland in harmony.

华东、华中、华南平原丘陵湿地区

华东、华中、华南平原丘陵湿地区北为淮河，东及东南为大海，西至巫山、雪峰山一线，南至我国热带北缘线。包括长江中下游平原、东南丘陵、东南沿海及沿海诸岛、台湾岛的北部，属中国地势第三级阶梯的中部分，气候区为亚热带。这里夏季高温多雨，冬季温和少雨，属亚热带季风气候。本区最重要的河流是长江和珠江，其干流均自第二级阶梯下来，并有大量密如蛛网的一、二级支流注入。另外，地区性河流如钱塘江、闽江、韩江等直接入海，还有人工湿地的京杭大运河南段。这里濒临海洋、气候温暖潮湿、水热条件优越，尤其是长江中下游平原和珠江三角洲地区，水网交织、河湖密布，水生动植物种类繁多、生长茂盛，是一片人与自然和谐共处的鱼米之乡。我国著名的淡水大湖——鄱阳湖、洞庭湖、洪泽湖、巢湖、太湖等都在本区。这里的河湖成因大多与河流水系有关，甚至和海涂的发育、海岸线的变迁有关，如著名的杭州西湖等。长江口以南的沿海以岩石性海滩为主，以北的沿海以淤泥质、砂质海滩为主。这里温暖湿润的环境产生了各种沼泽，沼泽的基本类型为淡水芦苇、苔草沼泽，泥炭层不发达，在丘陵山地由于潮湿多雾，有泥炭藓发育。由于气候适宜、水量充足，无霜期较长、水生动植物生长量大，每年秋冬季节湖泊水位低落，形成大面积的浅滩沼泽，吸引了种类繁多、数量庞大的水禽群体前来觅食、栖息、繁殖和越冬，是亚洲东部水禽中心和亚洲水禽南迁的必经之地。这里是白鹤、白枕鹤、白头鹤、东方白鹳等珍稀鸟类和各种雁鸭类集中的越冬地，台湾岛台南的滨海滩涂以及香港的米埔、深圳的福田等地是黑脸琵鹭的主要越冬地之一，在鄱阳湖越冬的白鹤数量占世界总数的95%以上。我国特有珍稀物种的扬子鳄、江豚等就生活在长江中下游流域一带。

在这个区域，人与湿地的关系尤其密切，它托起了中华文明的璀璨湿地文化，依靠着丰富的湿地资源，这片土地养育着中国超过1/4的人口。这里水网交织、河湖密布；阡陌纵横、舟楫便利；古镇散布，处处稻米花香，素有"人间天堂""鱼米之乡"及"湖广熟天下足"等的美誉。是千百年来人和湿地相互适应、互为因果、和谐相处和"师法自然"的典型。但是，当前水质污染和富营养化、水资源短缺、湿地萎缩等问题严重地影响着这里社会经济的和谐发展。

（从左至右）
钱塘江入海口 · 浙江杭州湾大桥
洪湖水清 · 湖北荆州
湖畔沼泽 · 江西鄱阳湖
野柳岩石海岸 · 台湾台北
水乡泽国话江南 · 江苏苏州
(From Left to Right)
The Estuary of the Qiantang River— Hangzhou Bay Bridge, Zhejiang
Cristal Clear Water in Honghu — Jinzhou, Hubei
Lakeshore Marshes — Poyang Lake, Jiangxi
Rocky Beaches at Yeliu — Taibei, Taiwan
A Village on the Water — Suzhou, Jiangsu

East, Central and South China Plains and Hilly Areas Wetland Zone

East, Central and South China Plains and Hilly Areas Wetland Zone is located in the area bordered on the north by Huaihe, east and southeast by East China Sea, west by Wu Mountain and Xuefeng Mountain, and south by the northern edge of China's tropical zone, including plains at the lower and middle reaches of the Yangtze River, foothills, coastal zones and offshore islands in Southeast China, and north part of Taiwan Island on the centre of the third step of China's terrain. This area belongs to sub-tropical monsoon wind climate, featuring in high temperature and rainfall in summer and mild winter with less rainfall. The Yangtze River and Pearl River, which receive innumerable first and second tributaries in this area after the main streams flow downstream from the second step of China's terrain, are the most important rivers. In addition, there are also local rivers such as Qiantang River, Min River and Han River that run directly into the sea, and south part of the constructed Grand Canal. Adjacent to the sea, this area enjoys warm and moist climate and superb water resources and sunlight. Especially the middle and lower reaches of the Yangtze River Plain and the Pearl River Delta, which are interweaved with water courses, rivers, lakes supplying abundant aquatic animals and plants, are lands flowing with milk where man and nature co-exist in harmony. All China's well-known freshwater lakes, including Poyang, Dongting, Hongze, Chao and Tai are found here. The cause of river and lake here is usually related to river system and to some extent the development of coastal tidal flat and the change in coastal lines, e.g. the famous West Lake in Hangzhou. Coasts to the south of the estuary of the Yangtze River are primarily rocky beaches, whilst muddy and sandy coasts are found primarily to the north. The mild and moist environment results in numerous marshes categorized in fresh reed marsh and carex marsh. Peat is underdeveloped and sphagnum grows in foothills due to the humid and foggy weather. Aquatic animals and plants grow quite well under favorable climate with abundant water supply and long frost-free period. Large areas of shallow flats and marshes occurred in autumns and winters when water table drops attract many waterfowls to come in flocks for forage, inhabiting, breeding and wintering. This area is the center of waterfowls in East Asia and a pathway which waterfowls in Asia must take when migrate to the south. It is the wintering site of rare and endangered birds such as Siberian crane, white-naped crane, hooded crane, oriental stork, and various ducks and gooses. The coastal tidal flats in Tainan of Taiwan Island, Mai Po in Hong Kong and Futian in Shenzhen are the main wintering sites of black-faced spoonbill. More than 95% of the world's Siberian crane population winters in Poyang Lake. Some rare and endangered species such as Chinese alligator and finless dolphin in China are found in the middle and lower reaches of the Yangtze River.

The relation between man and wetlands is particularly close in this area where the wetland culture of Chinese civilization develops and over one fourth of Chinese population feeds on rich resources from wetland. It is known as Heaven on Earth, A Land Flowing with Milk and "When Huguang has a bumper crop, the whole country will be free from hunger" for the interweaving water courses, rivers and lakes, convenient water transportation system, ancient villages and plenty of paddy fields. This is a good example of interaction and mutual adaption between wetland and man who always learns from nature. However recent problems such as water pollution, eutrophication, water shortage and wetland decrease are severely impeding the social and economic development in this area.

珠江口虎门大桥 · 广东虎门

珠江发源于第二阶梯的云南沾益马雄山，经贵州、广西、广东注入南海，其流域还涉及湖南、江西两省，是我国列在长江、黄河之后的第三长河流，年流量仅次于长江，是中国七大江河中含沙量最小的河流。

Humen Bridge at the Mouth of Pearl River — Humen, Guangdong

Originated from Maxiong Mountain at Zhanyi, Yunnan on the second step of China's terrain, the Pearl River flows downstream to Guizhou, Guangxi and empties into the South China Sea in Guangdong. Its watershed includes also Hunan and Jiangxi. It is the third longest river in China after the Yangtze River and the Yellow River, however its annual discharge volume ranks in the second place after the Yangtze River, and its sediment concentration is the smallest among seven major rivers in the nation.

繁忙的长江 · 湖北宜昌

三峡以东，地势豁然，江面骤然展宽，长江进入了中国地势的第三阶梯。长江中下游均位于该区，这里幅员辽阔，江湖众多，土地肥沃，气候温和，资源丰富，历史文化悠久，既是中华民族的重要发祥地，也是我国总体经济实力最为雄厚的地区。长江是连接中国东、中、西部水路运输的大动脉，是世界上运输最繁忙、运量最大的内河水道，它有力地促进了沿江社会经济的飞速发展，成为引领中国经济快速发展的重要引擎。

Busy Traffic on the Yangtze River — Yichang, Hubei

The surface of the Yangtze River abruptly turns wide eastward of the Three Gorges after it enters downstream to the third step of China's terrain. Spanning over a large territory, the middle and lower reaches of the Yangtze River is endowed with plentiful rivers and lakes, fertile soil, mild climate, rich resources, long history and culture features. This area is an important birthplace of the Chinese people and has the biggest economic power in the country. As the busiest inland waterway in the world with the biggest transportation volume, it is a major transportation artery connecting the east and central with the west, as well as a significant engine that boosts the rapid economic and social development both along the river and in the entire country.

日月潭晨光 · 台湾南投

日月潭是我国台湾最大的天然湖，由玉山和阿里山之间的断裂盆地积水而成，是全国著名的高山湖泊之一。

Sun Moon Lake at Dawn — Nantou, Taiwan

Formed by fault basin between Yu Mountain and Ali Mountain, Sun Moon Lake is the biggest natural lake in Taiwan, and one of the well-known highland lakes in the country.

漓江山水甲天下 · 广西阳朔

〝梦里仙境心意遐，水墨奇峰舟为家〞的喀斯特湿地地貌。

Landscape in Li River Tops the Country — Yangshuo, Guangxi

The Karst wetland landscape is described as "Living in a boat drifting in such a picturesque landscape is just like living in the dreamed fairyland".

德天瀑布光影 · 广西崇左

德天瀑布是跨于中越边境线上的瀑布，排在巴西 - 阿根廷之间的伊瓜苏瀑布、赞比亚 - 津巴布韦之间的维多利亚瀑布以及美国 - 加拿大的尼亚加拉瀑布之后，是世界第四、亚洲第一跨国瀑布。瀑布气势磅礴，一波三折，层层跌落，水势激荡，声闻数里。

Lights on Detian Falls — Chongzuo, Guangxi

Standing across the border of China and Viet Nam, Detian Falls is the 4th largest waterfall along a national border after Iguazu Falls (Brazil-Argentina), Victoria Falls (Zambia-Zimbabwe) and Niagara Falls (US-Canada), and the largest in Asia. This magnificent waterfall composed of three falls is famed for water volume in peak season and the thundering effect of the water hitting the cliffs that can be heard from afar.

西湖美景 · 杭州

西湖是一个潟湖，是中国最著名的观赏性淡水湖。远在秦朝时，西湖还是一个和钱塘江相连的海湾，后来由于潮汐冲击及泥沙淤积，使海湾和钱塘江分开来变成了内湖。西湖是目前中国列入《世界遗产名录》中唯一一处湖泊类文化遗产。

The Beauty of West Lake — Hangzhou

The West Lake, a lagoon, is China's most famous and spectacular freshwater lake. Originally part of a bay connected with Qiantang River back in the Qin Dynasty, it then was separated from the river and became an inland lake due to tidal impact and sedimentation. It is the only lake-type culture heritage in China that is inscribed on the UNESCO World Heritage List.

珍稀鸟类的天堂 · 江西鄱阳湖

鄱阳湖为中国第一大淡水湖。每年秋去冬来，水落滩出，沼泽星罗棋布，水草繁茂，鱼虾众多，这里是候鸟理想的越冬地，其中仅属国家一级保护的动物就有 9 种，属国家二级保护的动物有 32 种，全球 95% 以上的白鹤在此越冬，也是迄今发现的世界上最大的鸿雁越冬群体所在地。图为在鄱阳湖越冬的白鹤和白枕鹤群。

Paradise for Rare and Endangered Birds — Poyang Lake, Jiangxi

Poyang Lake is the biggest freshwater lake in China. Each winter when flat exposes from the shallow water and densely scattered marshes begin to supply abundant plants, fishes and shrimps, it becomes an ideal wintering site for migratory birds, which include 9 species under the 1st Grade National Protection, 32 under the 2nd Grade National Protection, and over 95% of the world's Siberian crane population. The biggest wintering swan goose population discovered in the world is also found here. The photo shows Siberian crane and white-napped crane population that winter in Poyang Lake.

水禽的重要栖息地 · 湖南洞庭湖

洞庭湖是一个宽阔河道型湖泊，它承纳长江中上游和湖南的湘、资、沅、澧四条江河的水量，呈现出一派江湖沼泽、平原河网的地貌景观，是中国湿地水禽的重要越冬地和重要的繁殖地、停歇地，每年在这里栖息的水鸟达数百万只。在包括洞庭湖、鄱阳湖等在内的长江中下游湿地中生存着 300 多种鸟类、600 多种水生、湿生植物和 400 多种鱼类。图为洞庭湖飞翔的鹤鹬群。

A Critical Habitat for Waterfowls — Dongting Lake, Hunan

Being a river-type lake that is fed by rivers in the upper and middle reaches of the Yangtze River and four major rivers (Xiang, Zi, Yuan, Li) from Hunan, Dongting Lake, with its innumerable marshes and river networks, is a critical wintering, breeding and stop-over site for China's waterfowls. Each year over millions of waterfowls inhabit in the lake. More than 300 birds, 600 aquatic and wetland plants and 400 fishes choose to live in the wetlands along the middle and lower reaches of the Yangtze River, including Dongting Lake and Poyang Lake. The photo shows flying spotted redshanks in Dongting Lake.

滩涂霞光 · 江苏盐城

这里是中国沿海最大的一块滩涂湿地，是古黄河三角洲和长江三角洲的泥沙在黄海和东海波浪冲撞及潮汐作用下形成的，为淤泥质平原海岸的典型代表。这里孕育着大量的生物，为数以百万计的水禽迁徙提供了栖息场所，更是濒危物种丹顶鹤的重要越冬地。

Sunglow on the Tidal Flats — Yancheng, Jiangsu

As the biggest tidal flat wetland along the coastal zone of China, this typical muddy coast is formed by the sedimentation of ancient Yellow River Delta and the Yangtze River Delta under tidal action and wave impact from the Yellow Sea and East China Sea. Endowed with plenty of organisms, it provides habitat for millions of migratory birds, and is most importantly a critical wintering site for the endangered red-crowned crane.

海上生明月 · 浙江洞头

农历八月十五，海上的月亮格外明亮，对沙滩温柔抚爱的潮水静静地退到了远远的地方，把船只和月亮一并都搁在了沙滩上。

A Round Moon above the Sea — Dongtou, Zhejiang

The moon appears brighter at a Chinese Moon Festival night. The tides that once gently fondled the sand retreat quietly, left the boat and the moon on the beach.

东海上大潮 · 浙江洞头

没有对沙滩温柔的抚爱，就没有对岩石猛烈的冲撞。涨潮时海浪冲到了它能够到达的最远地方，于是就在海岸上制造了最大的潮间带。

Flood tides of East China Sea — Dongtou, Zhejiang

Without tender love to the sand, there will be no crashes to the rocks. Waves crash to the furthest distance it could at high tides, thus create the biggest intertidal zone on the coasts.

1

2

3

4

5

6

1. 浮水植物荇菜 · 安徽池州
Nymphoides peltatum, a Floating Plant — Chizhou, Anhui

4. 水中"落花生"菱角 · 浙江宁波
Trapa bicornis, also known as water groundnut — Ningbo, Zhejiang

2. 挺水植物黄菖蒲 · 上海青浦
Iris pseudacorus, an Emergent Plant — Qingpu, Shanghai

5. 冬季开花的疏花水柏枝 · 重庆长江三峡
Myricaria laxifolia Which Blossoms in Winter— The Three Georges of the Yangtze River, Chongqing

3. 浮水植物浮萍 · 湖北荆州
Lemna minor, a Floating Plant — Jingzhou, Hubei

6. 沉水植物菹草 · 湖南益阳
Potamogeton crispus, a Submerged Plant — Yiyang, Hunan

梭鱼草 · 湖南岳阳

水生植物是指那些能够长期在水中正常生活的植物。根据水生植物的生活方式，一般将其分为挺水、浮水和沉水植物。它们常年生活在水中，形成了一套适应水生环境的本领，其突出特点是具有发达的通气组织，即使长期在不含氧气或氧气缺乏的污泥或水中，仍可以生存下来。通气组织还可以增加浮力，维持身体的平衡。在水生环境中还有种类众多的藻类，它们是很多生物如鱼、两栖、鸟类等的食料或繁殖的场所。

Pickerelweed — Yueyang, Hunan

Aquatic plants are those which can permanently grow in water. According to their different ways of growing, aquatic plants are generally divided into emergent plants, floating plants and submerged plants. Since they are growing in water all year round, aquatic plants have developed a set of skills which enable them to well adapt to the aquatic environment. One outstanding feature of aquatic plants is their well-developed aerenchyma, which helps plants to survive even in the absence of oxygen or when there is a lack of oxygen in the sludge or water. Aerenchyma can also increase the buoyancy ventilation to maintain the body balance of the plants. In the aquatic environment, there lives a wide variety of algae, which become food or breeding places for aquatic creatures, such as fish, amphibians, birds, and so on.

浮出水面呼吸的江豚 · 安徽铜陵

长江江豚被称作长江生态的“活化石”和“水中大熊猫”，目前数量已经远远小于大熊猫了。2001 年，中外科学家宣布白鳍豚功能性灭绝后，江豚就成了长江里唯一的淡水豚类。调查研究证明，白鳍豚灭绝是因为人类的活动，包括航运的发展、渔业的延伸和污染的排放造成的，如果江豚的生境仍然得不到根本性改变，未来 15 ~ 20 年江豚可能会重蹈白鳍豚的覆辙。

Finless Dolphin Catching Breath above the Water Body — Tongling, Anhui

Finless dolphin (*Neophocaena phocaenoides*) in the Yangtze River is also known as the "living fossil" and "giant panda in water". Its total number is currently far below the number of the giant panda. After the Chinese and foreign scientists declared in 2001 the functional extinction of Yangtze River dolphin (*Lipotes vexillifer*), finless dolphin has become the only freshwater dolphin species in the Yangtze River. Research shows that the extinction of Lipotes vexillifer has been caused by human activities, including the navigation and fish industries, as well as pollution emissions. If there will not be any significant changes to the habitat environment of finless dolphin, the tragedy of Yangtze River dolphin might happen again to finless dolphin in the next 15 to 20 years.

1

2

3

4

5

6

1. 丹顶鹤 · 江苏盐城
Red-crowned Crane — Yancheng, Jiangsu

2. 黑脸琵鹭 · 香港米埔
Black-faced Spoonbill — Mai Po, Hongkong

3. 麋鹿 · 江苏大丰
Milu (Père David's Deer) — Dafeng, Jiangsu

4. 白鹭 · 广东潮州
Little Egret — Chaozhou, Guangdong

5. 中华花龟 · 浙江温州
Chinese Striped-neck Turtle — Wenzhou, Zhejiang

6. 扬子鳄 · 安徽宣城
Chinese Alligator — Xuancheng, Anhui

白鹤亮翅图 · 江西鄱阳湖

在越冬地，白鹤往往在高兴、嬉戏时不断对天鸣叫，有时还同时将翅膀张开，表现出异常的兴奋状态。这时，鄱阳湖冬天的晨雾刚刚散开，鸟类天堂的一天又将开始了，白鹤群用自己兴奋的表现呈给我们一幅美丽的水墨淡彩中国画。

White crane spreads its wings — Poyang Lake, Jiangxi

In wintering grounds, cranes often sing when they are happy or enjoying play, sometimes they will open wings, showing the abnormal state of excitement. At this point, the morning winter mist in the Poyang lake just scattered, another day for the paradise of birds begins. The performance of the crane group with their excitement presents us a beautiful spectacle similar to a Chinese painting with ink and light color.

渐出舞台的东方白鹳 · 湖北沉湖
东方白鹳的越冬地主要集中在长江中下游的湿地湖泊，目前这里因围垦、泥沙淤积等原因，在一些曾经的越冬地中，东方白鹳的数目已经显著下降或根本消失。近来有些地方甚至发生毒杀现象，使本来就濒危的这一物种的生存更加雪上加霜。

Oriental White Stork Fading out of the Stage — Chen Lake, Hubei
Winter grounds for oriental white stork are mainly concentrated in the wetland lakes along the middle and lower reaches of the Yangtze river, where at present the number of oriental white stork have fallen significantly or disappear altogether due to land reclamation, sedimentation and other reasons. Recently poisoning has been spotted in some places which aggravate the hard survival of this already endangered species.

著名的江南水乡 · 上海朱家角

长江三角洲平原湖泊众多，河流纵横，大部分地区的光、热、水资源充沛，素有“鱼米之乡”之称。这里河湖交错、水网密布，小桥流水、古镇村落，是千百年来人与湿地和谐相处的“人间天堂”。

Famous Water Town at the South of the Yangtze River — Zhujiajiao, Shanghai

The Yangtze River Delta Plain has a large number of lakes and rivers. Most of the area has abundant light, heat and water resources, and is known as the "land of agriculture and fishery." This is a picturesque place because of the staggered rivers and lakes, dense water network, streams running under numerous bridges and ancient style towns and villages. This place has been a "paradise of the mankind" for the past thousands of years. Human beings and the wetlands exist in harmony.

江南水乡文化凝结地 · 浙江西溪

西溪湿地集生态湿地、城市湿地、文化湿地于一身，是人与湿地长期和谐相处的结果，这里建立了中国的第一个国家湿地公园。

A Town with Combination of Water and Culture at the South of the Yangtze River — Xixi, Zhejiang

Xixi Wetland has everything in itself – an ecological wetland, an urban wetland and a cultural wetland. It is an outcome from a long and harmonious relationship between mankind and the wetland. China's first national wetland park was established in this area.

中国经济腾飞的长江龙头 · 上海

发源于青藏高原，流经中国地势三级阶梯，十一个省份，长 6397 千米的长江，在这里汇集了它的最后一条支流——黄浦江，终于注入东海了。不论是以过去的外滩为代表，还是以现在的浦东为代表，上海都以她得天独厚的综合区位优势，引领着长江三角洲，以至长江流域的生产生活，为中国生产力最发达地区的龙头。

The Imperial Economic Center of China along the Yangtze River — Shanghai

The Yangtze River rises from the Qinghai-Tibet Plateau, flowing through the three steps of China's ladder topography and passing through eleven provinces, with a total length of 6,397 kilometers. It finally flows to the East China Sea from the last tributary Huangpu River. Either with the Bund as its representative in the past, or Pudong as its representative at present, Shanghai has always been the biggest development production force in China, thanks to its unique geographical advantage. It has been the imperial economic center in the Yangtze River Delta, and even in the entire Yangtze River catchment.

AURORA

6 西南、华南山地湿地区

Southwest and South China Mountainous Areas Wetland Zone

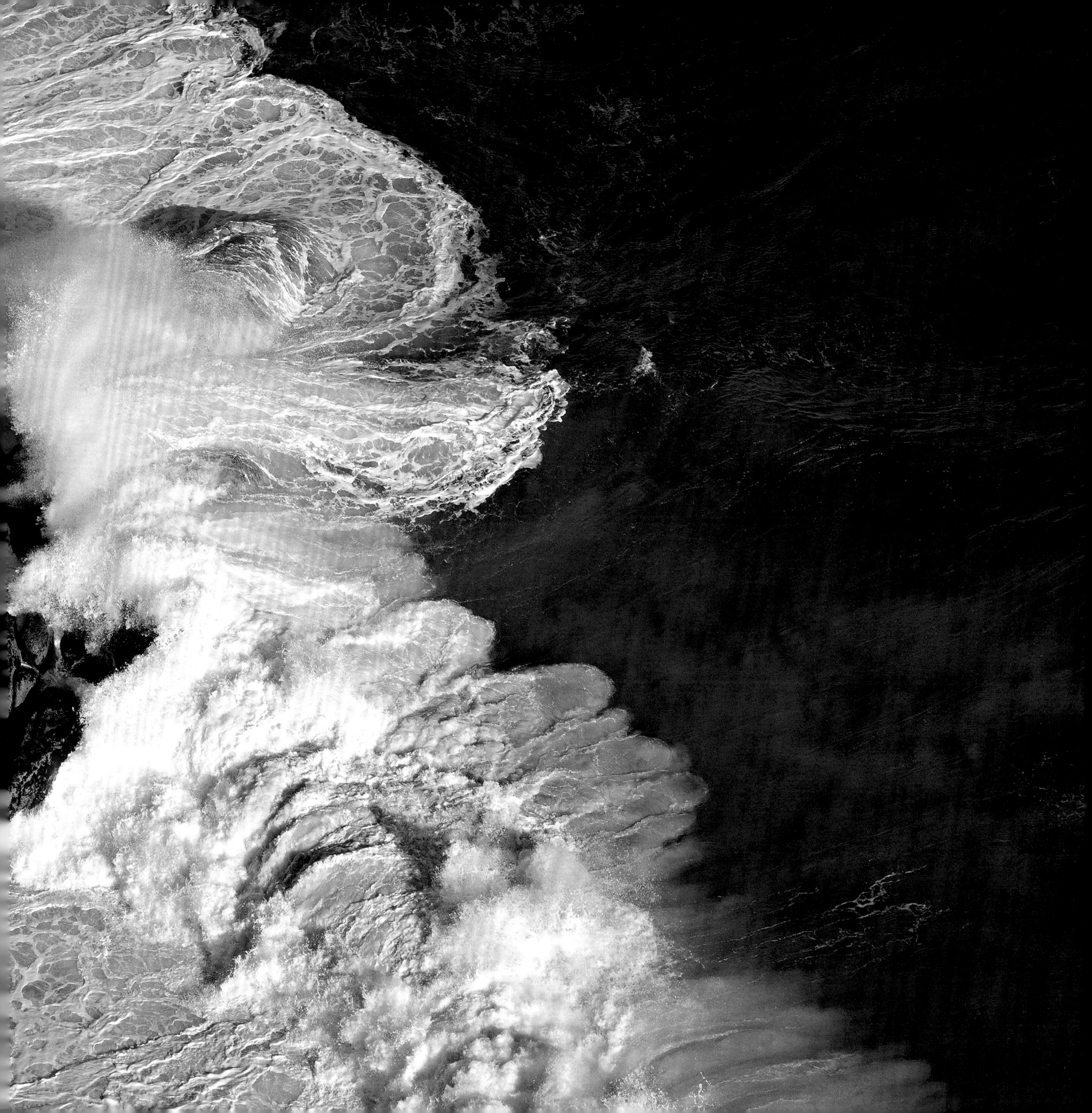

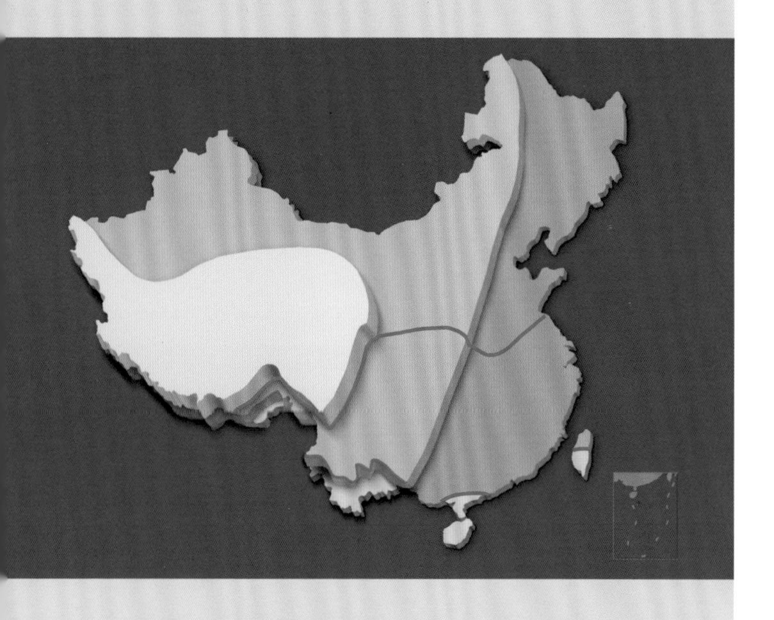

（前页）岩石海岸 · 台湾花莲

我国的基岩海岸多由坚硬岩石组成，主要分布于杭州湾以南的漫长海岸线上，还有山东、辽东半岛的海岸和台湾东海岸。基岩海岸最为壮观美丽的景象是从海上奔腾而来的巨浪犹如大块翡翠撞击在岩石上，即刻就被摔得粉碎，白色的浪花四起，气势磅礴，发出阵阵轰鸣，具有变幻无穷、雷霆万钧的气势和神韵。

(Previous Page) Rocky Seacoast— Hualian, Taiwan

Most of the bedrock coast in China is made of hard rocks and is mainly distributed in the south of the long coastline along the Hangzhou Bay, and some in the coastlines along the Shandong Peninsula and Liaodong Peninsula, and the east coast of Taiwan. The most spectacular sight of the bedrock coasts is the moment when the huge sea waves, which look like a giant jade, crash on the rocks and instantly are broken into pieces. The foaming waves rise everywhere, making people feel its magnificent and ever changing charm, and its strong power, which can be as enormous as thunderstorms.

西南、华南山地湿地区

西南、华南山地湿地区是我国第三阶梯的南部分暨热带部分，即热带与亚热带分界线以南的所有地区，包括台湾、广东、云南的最南部及藏东南的最南部，海南全部（包括南海诸岛），这些地区多为山地。漫长的海岸线上岩石、沙石、淤泥质海滩相互交叉分布，而南海中多为珊瑚岛礁、环礁和潟湖等。该地区属热带季风气候，特点为长年高温，年降水量大，旱雨季明显。这里除云南南部有澜沧江、元江等大河从第二阶梯的云贵高原下来并过境外，其余河流都是源于山地直接入海的短河流，如台湾的曾溪、高屏溪，海南的南渡江、昌化江、万泉河、陵水等。由于山地河短，很少有湖泊产生。在高温高湿的环境里尽管温度、湿度有利于生物生长，但由于气温太高，分解过程占据优势，泥炭积累受到抑制，内陆地上不易形成沼泽，在沿海红树林的特殊环境下却容易形成沼泽。我国的红树林主要就分布在这个地区，红树林沼泽与大陆迥异的气候条件、宽阔的潮间带滩涂以及丰产的甲壳类动物和水生植物吸引了成群的海鸥、鸻鹬类、鹭类、雁鸭类等在这里栖息、繁殖和越冬。台湾岛台南的滨海滩涂是黑脸琵鹭的主要越冬地，种群数量较大。南中国海的诸多珊瑚礁盘、环礁、潟湖、暗礁等是珊瑚、藻类、甲壳类、鱼类等水生生物最好的栖息繁衍地，西沙群岛是红脚鲣鸟的唯一繁殖地。总之，这里湿地的组成与其他地方不同的是，河流出山入海很短，几乎没有沼泽（除红树林沼泽外）和湖泊，滨海湿地及海洋中的环礁、潟湖湿地所占湿地的比重很大。

这里湿地的组成不同，人们在处理与湿地的关系也就大大不同，与传统的农耕文化不同的海洋文化是这里的主打。人与海、中华民族与海的关系在这里显得尤其重要，现在重要、将来会越来越重要。我们必须坚定地、尽快地走向蔚蓝色的大海，强化对包括海南诸岛屿、环礁、潟湖等在内的湿地的保护和利用，这个重大战略布局必须引起我们国家、民族的高度重视。

（从左至右）
干热河谷 · 云南红河
水平如镜山尽收 · 海南尖峰岭
红树林 · 广西山口
海岛卫士 · 海南永兴岛
请到天涯海角来 · 海南三亚

(From Left To Right)
Dry and Hot River Valleys— Hong River, Yunnan
As Flat as a Mirror and an Amazing View of the Mountains — Jianfengling Mountain, Hainan
Mangrove Forest— Shankou, Guangxi
Island Guardian— Yongxing Island , Hainan
Welcome to the Furthest Ends of the Earth— Sanya, Hainan

Southwest and South China Mountainous Areas Wetland Zone

Southwest and South China Mountainous Areas Wetland Zone is located at the south part of the third step of China's terrain and fall to the tropical zone climate category. Namely, they include all the areas south of the tropical and sub-tropical dividing line, which range from the most southern part of Taiwan, Guangdong, Yunnan and southeast Tibet, to the entire Hainan Province (including the islands in the South China Sea). Most of the areas here are mountainous lands. rocks, sand, silt beach spread across long coastlines, while in the South China Sea the dominant wetlands are coral reefs, atolls and lagoons. The area falls to a tropical monsoon climate category, with year-round high temperature, abundant annual rainfall, and distinguishable dry and raining seasons. Among the rivers flowing through this region, except the southern part of Yunnan Province where some big rivers, such as Lancang River and Yuanjiang River, rise from the Yunnan-Guizhou Plateau, which is the second step of China's terrain, all other rivers are short in length because they rise from rolling lands and enter directly to the sea. These short rivers include Taiwan's Zengxi and Gaoping, and the rivers of Nandu, Changhua, Wanquan and Lingshui in Hainan Province. Because of the short length of the rivers, lakes are rarely formed and found in this area. Although the temperature and humidity under a high temperature and high humidity environment is conducive to biological growth, the extreme high temperature will enable the decomposition process to occupy a dominated role, therefore, inhibit peat accumulation, and thus makes it difficult for inland marshes to formulate and grow. Instead, it is easy for swamps to develop under special circumstances where coastal mangroves exist. China's mangrove forest is primarily distributed in this area. Because mangroves swamps have a vast contrast of climate conditions from that of the territorial continental, plus its wide intertidal mudflats as well as the rich crustaceans and aquatic plants, this area has become an attractive place for flocks of gulls, waders, herons, Anatidae birds to live, breed and winter. The coastline at the south of Tainan is the major wintering place for Platalea minor with large population. Many round coral reefs, atoll, lagoons and submerged reefs in the South China Sea are the best breeding and living places for aquatic organisms such as coral reefs, algae, crustaceans and fish. The Paracel Islands is the only breeding place for red-footed boobies. The difference of the wetland composition here from that in other places is that the distance from the mountains where the rivers rise to the place where the rivers enter the sea is very short, hence almost no swamps (except mangrove swamps) and lakes exist. Also, there is a very high proportion of marginal coastal wetlands, atolls and lagoons.

Because of the different wetland composition here, people's attitude towards wetlands is much different accordingly. Compared to the traditional farming culture, marine culture dominates in this area. The relationship between mankind and the sea, and between the Chinese nation and the sea, is particularly important and will become increasingly so in the future. We must act firmly and quickly to strengthen the protection and utilization of the wetlands, including the islands, atolls and lagoons in South China Sea. This is an important strategy to which the government must pay high attention.

鉴江入海口 · 广东湛江

图为广东鉴江入海口，前面是河流与潮汐共同作用下自然形成的海沙垄（即拦门沙）。滨海湿地是人类文明的发祥地之一，如今已成为全球发展最快、人口密度最大的地区，世界大约 1/3 的城市人口居住在离海岸 60 千米以内的范围。

Estuary of Jianjiang River — Zhanjiang, Guangdong

This is Jianjiang River estuary in Guangdong Province. In front of the estuary lies the tidal sand ridge (i.e. sandbar), which has been formed naturally under the joint force of the river and the tides. The costal wetlands are one of the cradles of human civilization and has now become the world's fastest growing and most densely populated areas. About one third of the urban population lives on lands within a parameter of less than 60 kilometers away from the seacoast.

昌化江之源 · 海南五指山

海南岛以五指山、霸王岭等为主的山脉，是海南主要河流昌化江、万泉河、南渡江、陵水的发源地，主宰着海南岛的水系生态。因降雨量充沛，山地河短形成了很多瀑布，且往往都隐藏在葱翠茂盛的热带植被中，为本地区的一大特色。

The source of ChangHua River — Wuzhi Mountain, Hainan

Wuzhi Mountain and Bawangling mountains in Hainan are the birthplace for Hainan's major rivers including ChangHua river, Wanquan river, Nandu river, and Lingshui river which dominate the river ecology in Hainan island. Due to abundant rainfall, mountainous landscape and short rivers, many waterfalls are formed which usually hide in a lush tropical vegetation. This is one of the features of the region.

南海渔港晨光 · 广东徐闻

中国热带的海就是南海，这里海阔滩广，海岸近处的小渔船、远处的大渔轮呈现的是一片繁忙景象，中国南海渔民世世代代就生活在这里。中国的大陆海岸线绵延 18000 多千米，有 6500 多个大小岛屿，岛屿海岸线 1400 多千米，海岸线总长度居世界第三位，滨海湿地面积约 594 万公顷（未包括南海），2 万多种海洋生物在这里栖息繁衍，海洋资源非常丰富。

Dawn in the South China Sea Port— Xuwen, Guangdong

The South China Sea is the only tropical sea in China. It has wide seacoast and presents people a scene of busy life – everywhere you can find small fishing boats close to the shore and large fishing vessels at the far end of the sea. This is the place where fishermen have been living for generations. China's continental coastline stretches over a length of 18,000 kilometers, with approximately 6,500 small islands and more than 1,400 kilometers of island coastline, which ranks the third in the world. The total area of costal wetland (excluding the South China Sea) is about 5.94 million hectares with abundant marine resources and inhabits more than 20,000 kinds of marine organisms.

赶海 · 海南文昌
休渔期间，渔船都静静地躺进了港湾的怀抱，闲暇的渔民趁潮退时机搜寻一些贝蟹鱼虾，谓之“赶海”。

Beach Combing — Wenchang, Hainan
During the moratorium, fishing boats lay quietly in the bay. Fishermen become beach combers, who take advantage of the low tide at leisure to find and collect shellfish, crabs and fish.

条沟状海滩 · 广东沃内海
由于地理、潮流、方位等综合因素，这里的沙滩呈罕见的条带状。退潮后沙滩上到处是招潮蟹从地窝里推出的泥球。

Ditch-shape Beaches— Wonei Sea, Guangdong
Due to influence of many factors such as geography, tidal current and locality etc., the beach has been formed in the shapes of ditches. At low tidal hours, the beach is full of mud balls pushed out by fiddler crabs (Uca.) from their underground caves.

五彩的海滨 · 广东湛江
日升日降、潮涨潮落，船海交集，赋予海滩的是无穷变幻的色彩、周而复始的活力和绚丽多彩的生命。

Colorful Seacoasts — Zhanjiang, Guangdong
Sunset and sunrise, tidal recessions and returns, and massive number of ships on the sea have provided the beach with ever-changing colours and sustained resilience.

美丽的琛航岛 · 海南西沙

这是南中国海中无数美丽岛屿中普普通通的一个。蔚蓝色的大海拥抱着绿色葱郁的小岛，千百年来，珊瑚在海水作用下，尸骨堆砌在海岸边，形成了浅海、沙滩、潟湖和环礁。海水的深浅变化使海水从深蓝过渡到蔚蓝、淡蓝、淡绿，甚至到无色透明，这里是各种海洋生物栖息的绝妙天堂。

Beautiful Chenhang Island— Paracel Islands, Hainan

This is a common island among numerous beautiful ones in the South China Sea. The deep blue sea is surrounded by lush green islands. Over thousands of years in the past, impacted by the sea water, the coral skeletons are piled up on the coasts and enabled the formulation of shallow sea, beaches, lagoons and atolls. The change of sea water depth has brought a colour transition of the sea from dark blue, to light blue, light green, even to crystal clear. This area has become a wonderful paradise for a great variety of marine habitats.

珊瑚潟湖 · 海南西沙

珊瑚潟湖由环状珊瑚礁或由坝状珊瑚礁相隔而成，水域呈不规则形状，潟湖通海，但水都不深，是海洋生物生长繁育的极好环境。潟湖和环礁都是我国重要的湿地类型之一，这种类型在南中国海非常普遍，我们断不可遗漏。

Coral Lagoon— Paracel Islands, Hainan

Coral lagoons are waters enclosed by round coral reefs or dam-shape coral reefs. The waters are irregular in shapes. They are not deep and are connected to the sea, and are excellent breeding and growing places for marine organisms. Lagoons and atolls are one of the important wetland types in China. We must not ignore them because they are very popular in the South China Sea.

1

2

3

4

5

6

1. 红树的呼吸根
Pneumatophore of the Mangroves

2. 红树林中的藤蔓
Tree Vines in the Mangroves

3. 红树的板根
Buttress Roots of the Mangroves

4. 红树林中的招潮蟹
Uca in Mangrove Forests

5. 红树的胎生现象
Vivipary of Mangroves

6. "小"螃蟹撞上了"大"弹涂鱼
"Small" Crab encounters "Big" Mudskipper

红树林 · 海南东寨港

热带和亚热带海岸、海湾或河流出口处的盐沙壤土，比较适于红树林的生长和扩展。中国不到 220 平方千米的红树林中，就生活着超过 2300 种的海洋生物，其单位面积的物种是海洋平均水平的 500 倍。在 60 多年之前，中国红树林面积曾达 25 万公顷，而如今，91% 以上的红树林已经消失。红树林是滨海湿地最珍贵的形态之一，红树林的减少不仅是底栖动物和海洋生物的灾难，更是人类面临的灾难。

Mangrove Forest — Dongzhaigang, Hainan

Mangroves grow well on salt sandy lands in tropical and subtropical coasts, as well as at the sea bays or river estuaries. China's mangrove forest has an area of less than 220 square kilometers. It hosts more than 2,300 marine organisms. The unit number of species is 500 times the average level of the ocean. 60 years ago, the mangrove area in China was 250,000 hectares, while at present more than 91% of mangroves have disappeared. Mangrove forest is one of the most precious wetland ecology types. The decline of mangrove forest area is not only a catastrophe for benthic creatures and marine organisms, it is also a disaster to the human beings.

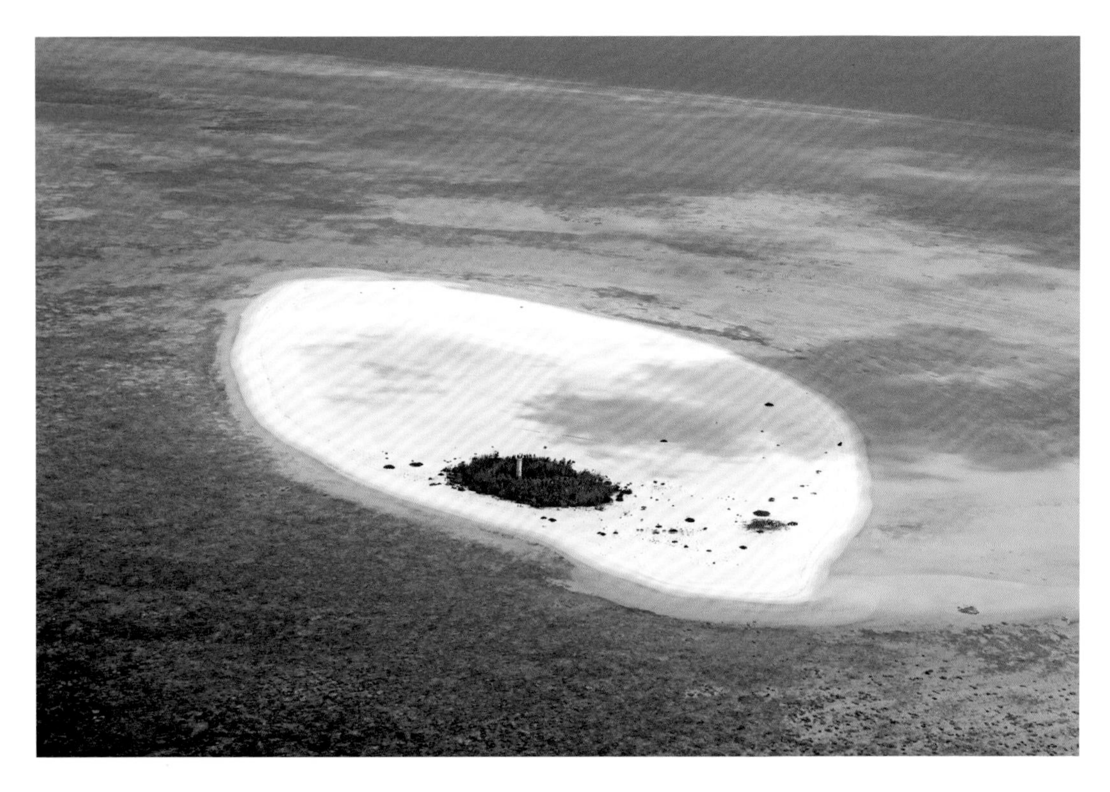

俯瞰西沙洲 · 海南西沙

位于南中国海西沙群岛七连屿最西侧的岛屿，名为西沙洲。西沙洲是一座正在扩大中的沙洲，洲呈椭圆形，大部为白色沙滩，沙滩逐渐延伸至水下庞大的礁盘上，礁盘上遍布艳丽多姿的各种鹿角珊瑚丛，异常美丽。该岛最高处有小片绿洲并建有灯塔，有丰富的淡水资源，是海龟上岸产蛋的主要岛屿之一。

A Bird-eye View of Xi Sha Zhou - Xisha, Hainan

Xi Sha Zhou is located at the west end of Qilian Islets (Seven Islets) of Paracel Islands in South China Sea. It is a growing oval sandbar covered mostly by white beach. The beach extends to the giant reef under the water which is full of colorful antler corals, presenting extraordinarily beautiful scenery. A light house was built on the small oasis at the peak of the island which maintains ample fresh water resources. Xi Sha Zhou is also one of the main islands for sea turtles to lay eggs.

翱翔在海云中 · 台湾台东

岩石海岸在海浪的不断撞击、涤荡下，海水的蓝、浪花的白逐渐相互交融在一起，汹涌澎湃而气势非凡。海鸬鹚在海天之间飞翔而过，仿佛置身于天上翻滚的云海之中。

Flying between the Clouds and the Sea — Taidong, Taiwan

Being slammed and washed continuously by sea waves, the rocky coasts gradually mingle together with blue sea and white waves, giving people an extraordinary momentum. Cormorants are flying between the sea and the sky. It seems that these birds are amid the rolling sea waters and clouds.

飞鱼 · 海南三沙
美丽的西沙、中沙、南沙群岛主要由珊瑚组成，浅海部分的生物资源相当丰富，飞鱼是生活在海水上层的鱼类，常常不知何时何处钻出水面并快速在空中飞翔一段，镜头极难捕捉住飞鱼突然出现并快速展翅飞翔的画面。

Skipjack — Sasha, Hainan
The beautiful Paracel Islands, Macclesfield Islands and Spratly Islands are primarily made of corals. There are abundant biological resources in the shallow sea waters. Skipjack is a kind of fish which lives on the upper section of sea waters. Very often skipjack will suddenly fly out of the water surface and make a turn in the air so quickly that it is extremely difficult for cameras to capture the moment.

1

2

3

4

1. 红脚鲣鸟 · 海南西沙群岛东岛
Sula Sula — East Island of the Paracel Islands, Hainan

2. 热带海鱼 · 海南南沙美济礁
Tropical Fish — Mischief Reef of the Spratly Islands, Hainan

3. 玳瑁 · 海南三沙
Hawksbill Turtle — Sansha, Hainan

4. 鲎 · 广东雷州
Horseshoe Crab — Leizhou, Guangzhou

银鸥的世界 · 海南三沙
银鸥是鸥科海鸟的一种，为候鸟。我国的台湾和海南（包括西沙、中沙、南沙群岛）是海鸥的重要越冬地。

World of Herring Gulls — Sansha, Hainan
Herring gull is a kind of sea bird of Laridea family, which are migratory birds. Taiwan and Hainan (including Paracel Islands, Macclesfield Islands and Spratly Islands) of China is the important wintering places for seagulls..

二代高脚屋 · 海南南沙赤瓜礁
这些建设在南海礁盘上的第二代高脚屋，饱含了在高温、高碱、高湿条件下守岛战士的艰辛，也体现了祖国海疆神圣不可侵犯的坚强意志。

Second Generation Rumah Buildings — The Johnson South Reef of the Spratly Islands, Hainan
These second generation rumah buildings constructed on the South China Sea reefs is not only a reminder of the hardship of the soldiers who lived here to safeguard the territory borders under harsh conditions such as high temperature, high alkali and high humidity, but also reflects the strong determination to protect the territory from any external invasions.

盐田湿地 · 台湾台南
海盐是当地人民生活不可或缺的必需品，在沿海一带盐田生产的海盐除了供人民生活需要外，还是重要的化工原料。

Salt Pan Wetlands — Tainan, Taiwan
Sea salt is an indispensable daily commodity to the local people. In addition, sea salt produced in the coastal area is also used as an important chemical raw material.

生态思考

Ecological Reflections

中国是世界上湿地类型最多的国家，也是湿地资源最丰富的国家之一。为了保护和利用好湿地资源，中国政府做出了巨大的努力。尤其是改革开放以来，中国的湿地保护翻开了新的篇章，1992年中国加入了《湿地公约》，2000年国务院17个部门联合颁布了《中国湿地保护行动计划》。国务院批准了《全国湿地保护工程规划（2002～2030年）》以及该规划的“十一五”和“十二五”实施方案。2004年，国务院办公厅发出了《关于加强湿地保护管理的通知》，明确了加强湿地保护是各级政府的重要职责，2013年国家林业局发布的《湿地保护管理规定》开始施行，“保护优先、科学恢复、合理利用、持续发展”是我国湿地保护管理的基本原则。这些年来，中国在湿地保护、湿地恢复、可持续利用示范和能力建设等方面都取得了重大的进展。

但是，由于长期以来人们生产、生活的需求对湿地资源造成的巨大压力和破坏，尤其是近几十年来我国社会经济的高速发展以及全球气候变暖、区域生态环境的变化，中国的湿地正面临着越来越严峻的挑战和威胁。主要表现在：对湿地的大量围垦和开发，对湿地生物资源的过度利用，水资源的不合理使用，湿地萎缩和污染日趋严重，气候变暖、冰川消融、干旱缺水、外来物种入侵、生物多样性锐减等等，这些都是对我国湿地资源以及生态环境的严重威胁和极大破坏。尤其是当今水资源问题越来越严峻地摆在国人面前的形势下，湿地保护更是关乎民生的安危，关系到我国社会经济能否可持续发展的大问题。

联合国在本世纪初作出的《千年生态系统评估》中强调，“近几十年来，湿地消失和退化的速度，超过了其他任何生态系统”。相对于其他生态系统来说，水陆交杂的湿地生态环境其生态构成关系更为复杂，哪怕是其中极小的变动，也会牵动整个生物链、整个生态系统的变化。湿地生态系统比其他生态系统来讲更加脆弱、更为敏感、更加复杂。主要体现在：

每块湿地并非一地一水独立存在，而是与其他湿地有着强烈的相互关联、相互影响的关系。上下游之间、流域之间体现出来的全局性、系统性都极强，各湿地之间可谓“一损俱损”。一个地方的湿地变化，不仅会敏感地影响该地区的水文、温度、湿度等小气候的变化，也会因为水分循环的系统关联性而影响其他地方的湿地和气候，可谓“牵一发而动全身”。

候鸟、鱼类等物种的迁徙将全球生态链编织成一个巨大的有机体，而分布于世界各地良好的一块块湿地组成的网络则是它们能否成功迁徙的重要保证。全球湿地网络中每一个湿地环境的微小改变，都有可能给迁徙物种带来致命的威胁。某一个地方的生态恶化，就会打断这个网链，直接影响到物种的迁徙或繁育，甚至引起某些物种的灭亡。一个物种的生存，又影响到与之相关的各种生物的生存，继而带来生物链的“多米诺骨牌效应”。生物链的变化最终会威胁到人类的生存以及整个地球，而我们“只有一个地球！”

近十年来,我国湿地面积在原来大大减少的基础上又减少了339.63万公顷（其中自然湿地面积减少了337.62万公顷），减少率为9.33%。国家重点生态功能区、湿地候鸟迁飞路线、重要江河源头、生态脆弱区和敏感区等范围内的重要湿地，还未全部纳入保护体系之中。全国湿地保护的空缺还较多，湿地保护管理任务非常艰巨。如：国家重点生态功能区湿地保护率仅为51.52%，国家重要湿地保护率仅为66.52%。威胁湿地生态状况的主要因子已从十年前的污染、围垦和非法渔猎三大因子，转变为现在的污染、过度捕捞和采集、围垦、外来物种入侵和基建占用五大因子。按照党的十八大提出的要“扩大湿地面积”的要求，国家林业局出台的《推进生态文明建设规划纲要》中划定了湿地保护红线，到2020年，中国湿地面积不得少于8亿亩。划定这条红线是遏制我国湿地资源面积减少，功能退化趋势的一个迫切需要。

湿地是重要的自然资源和具有独特功能的生态系统，保护湿地、维护湿地生态功能的正常发挥、科学管理和合理利用湿地，对于改善我国生态环境、确保水资源安全、促进社会经济可持续发展和建设生态文明社会都具有十分重大的意义。思考和应对这些问题是我们当代人不可推卸的历史重担，而解决这些问题更是我们以及今后几代人、十几代人都必须要承担的历史责任。

China is one of the countries in the world with richest wetland resources, and has the biggest number of wetland types. The Chinese government has made tremendous efforts in protecting and making good use of wetland resources. China's wetland conservation entered into a new chapter in particular when the government started the policy of reform and opening-up to the outside world. In 1992, China joined the *Ramsar Convention*. In 2000, 17 ministries and administrations of the State Council of China jointly issued the "*China Wetland Conservation Action Plan*". The State Council approved the "*China National Wetland Conservation Action Plan (2002-2030)*" and its associated implementation schemes for the 11th and 12th five-year plan periods. In 2004 the General Office of the State Council issued the *Notice on Strengthening Wetland Conservation*, which clearly pointed out that the responsibility of enhancing wetland conservation falls to governments at all levels. The implementation of the *Wetland Conservation Management Rules*, issued by the State Forestry Administration, was started in 2013. The basic principles in wetland protection in China are designed to give priority to conservation, to restore wetland in a scientific manner, to utilize wetlands wisely and to achieve sustainable development. Over the years, China has achieved significant progress in wetland conservation, rehabilitation, demonstration of sustainable use and capacity building.

However, China's wetlands are facing with more and more challenges and threats because of enormous pressure and destruction brought by the process of production activities and people's daily living needs. Other important factors include the rapid social and economy development China, the global warming and the regional ecological environmental changes in the past few decades. Specific examples, which form tremendous threats and enormous damage to wetland resources and ecological environment in China, include: excessive wetland reclamation, exploitation and irrational use of water resources, increasing wetland pollution, global warming, glaciers melting, drought, and invasion of exotic species and decline of biodiversity. Especially when we are facing an increasingly severe water problem, wetland is essential to secure people's livelihood and the sustainable development of China's social economy.

The *Millennium Ecosystem Assessment* of the United Nations stresses that wetlands degrade and become extinct at a far more rapid rate than any other ecosystems. Compared with other ecosystems, the structure of a wetland ecosystem with a mixed composition of water and land is more complex. Even one minor change can affect the entire food chain and cause changes to the entire ecosystem. The wetland ecosystem is more fragile, sensitive and complex than other eco-systems because of the following factors:

A wetland is not a pure piece of land and water which can function independently. Instead one individual wetland has strong interrelations with other wetlands and influences with each other. The overall integrity and systematic structure of wetlands, between upper reaches and lower reaches, among catchment areas and among various wetlands are so sensitive and strong that if one section becomes wrong, the entire system will be affected. Any wetland changes in one single locality will not only bring changes to the local microclimate in the aspects of hydrology, temperature and humidity, but also influence the climate and wetland in other regions because of systematic relevance of water circulation, just like the "butterfly effect".

The migrating activities of birds and fish make the global ecology chain an entire organism, while each well functional wetland across the world which forms a global network, becomes an important factor in the successful migration of birds and fish. In the wetland network, any minor changes in a single wetland environment may bring fatal threats to migratory species. Any deteriorations of ecology in one particular place will affect the overall chain, directly disturbing the migration or breeding of the species, and even causing extinction of some species. The existence of one species also affects the existence of other concerned species, and eventually may bring a "domino effect" on the entire food chain. Changes in the entire food chain will ultimately threaten the survival of humanity and the entire planet, and we have "only one Earth!"

Over the last decade, China's wetlands have witnessed a sharp decrease on the already downward trend by 3.3963 million hectares, or 9.33% of the total wetland area, 3.3762 million hectares of which is natural wetlands. Some important wetlands in national key ecological function area, flyways for migratory birds, major river source regions, and fragile and sensitive ecological areas have not been included into the protection system. There are still big gaps for wetland conservation across the country and the tasks remain arduous. For instance, only 51.52% of wetlands in the national key ecological function area are conserved and only 66.52% of national important wetlands are protected. The main factors that threaten wetland's ecology have shifted from pollution, land reclamation and illegal fishing a decade ago to the current pollution, over fishing, land reclamation, invasive alien species and infrastructure development. In line with the requirement of "widening wetlands area" upheld by the 18th National Congress of the Communist Party of China, the SFA has declared the red line for wetlands conservation in the Plan for Promoting Ecological Civilization Development, which stipulated that by 2020, the area of wetlands should be maintained at a level of no less than 800 million mu. The red line reflects the urgent necessity for checking the reduction and degradation of wetlands.

Wetlands are important natural resources and an ecosystem with unique functions. The protection of wetlands, ensuring the sufficient functioning of wetlands and scientific management and wise use of wetlands are extremely important for improving China's ecological environment, securing water resources, promoting sustainable social and economic development and building a society with ecological civilization. It is the inescapable responsibility of our generation to well consider and address the existing issues. At the same time, it is also a historical responsibility for our generation and the future generations to solve these problems.

威胁和挑战：
Threats and Challenges:

气候变暖　冰川消融

由于全球气候变暖和温室效应，地球上的冰川目前正在不断消融，速度也在进一步的加快。冰川是地球上最大的淡水库，全球70%的淡水被储存在冰川中。冰川的退缩或者消亡，冰川下游的河流也将干涸，湿地之源将遭受灭顶之灾，带来的是难以估量的生态灾难。据统计，青藏高原冰川末端在1976～2006年平均退缩速度为每年5米左右，2004～2006年退缩速度达到每年7.8米，并表现出近期加速后退的态势。高寒草甸正以每年上万亩的速度消失，荒漠化以每年几万平方米的速度扩大，迅速蔓延的荒漠化又加剧了水土流失，每年三江源区新增水土流失面积达21平方千米之多。图为川西曾经的冰川雪岭已经荒漠化了，这里位于四川雪宝顶自然保护区，中国最东的冰川就分布在这里。

Global Warming and Glacial Ablation

Due to global warming and the greenhouse effect, the earth's glaciers are currently melting at an increasing speed. Glaciers are the largest reservoir of fresh water on earth, storing 70% of global fresh water. Glacial recession or ablation will dry up the downstream rivers, and will leading to the extinction of wetlands sources. The ecological disaster brought about by these changes is beyond calculation. According to statistics, the glacier terminus of Qinghai-Tibet plateau retreated at an average rate of 5 meters per year from 1976 to 2006, and 7.8 meters per year from 2004 to 2006, and the retreat rate is showing acceleration recently. Alpine meadow are disappearing at an annual rate of nearly ten thousand mu, desertification expanding at an annual rate of tens of thousands of square meters, which is also aggravating soil and water erosion, and new soil and water erosion area in three-river headwaters region reached up to 21 square kilometers every year. This photo shows that previous glaciers and Snow Mountains in western Sichuan are desertified now. This place is located in the Xuebaoding Nature Reserve in Sichuan, where China's most eastern glaciers distribute.

极端灾害性天气频发

近年来，随着全球极端天气的加剧和自然环境的持续恶化，我国的水旱、地震灾害活动频繁，暴雪冻雨灾害损失重大，雾霾污染天气光顾不断，常常发出"50年一遇"甚至"百年未遇"的警告。湿地生态系统受到了前所未有的威胁和破坏，仅2010年洪涝灾害因灾直接经济损失就为3505亿元，是本世纪之最。图为四川汶川某地洪水带来的山体滑坡、河流改道、村庄被毁。

Frequent Extremely Disastrous Weather

In recent years, with the intensification of global extreme weather, and the worsening of natural environment, China's floods, drought and earthquakes occured blizzard and sleet brought heavy loss, haze and fog happened constantly. Warnings like "once in 50 years" or "once in a century" has been heard often. Wetland ecosystem has received the unprecedented threat and destruction. In the year of 2010, direct economic loss of RMB 350.5 billion was caused as a result of flooding, which was the largest amount of this century. The photo shows landslides, river diversion and destroyed villages caused by Wenchuan flood in Sichuan.

干旱缺水 资源短缺

中国是一个干旱缺水严重的国家，人均水资源仅为世界平均水平的1/4，在世界上名列121位，是全球13个人均水资源最贫乏的国家之一。如此短缺的水资源在时空上又分布不均。干旱缺水引发了湿地萎缩、草场退化、土地沙化、地下水位下降、生物多样性锐减等生态危机。这些年来中国的黄河、淮河等大河多次断流，一些地方性的河流在消失，数百个湖泊正在干涸，湿地在萎缩。图为几乎见底的华北某水库。

Drought and Resources Shortage

China is a country with serious problems of drought and water shortage, and its water resources per capita is only 1/4 of the world average, ranking the 121st in the world, and is one of the 13 countries with poorest water resources per capita. Moreover such deficient water resources distribute unevenly in time and space. Drought leads to a series of ecological crises including wetland decrease, grassland degradation, land desertification and the decline of the ground water level and sharp drop in biodiversity. In recent years, China's major rivers including the Yellow River, Huaihe river dry up for many times, some local rivers and streams are disappearing, and hundreds of lakes are running dry. The photo shows a dried up reservoir in northern China almost visible of the bottom.

水资源不合理利用

在水资源使用方面浪费无序，到处截留水源、大水漫灌庄稼、大量抽取地下水、污水回灌乱放等水资源不合理利用现象相当严重。致使很多湿地从此消亡，农地弃耕、林木死亡、地下水漏斗越来越大、小气候异常、地下水污染等等。当前，水危机已然成为最大的资源危机。全球气候变化加剧了中国的水危机，水资源与水灾害的格局将发生重大变化，不仅水、旱灾害频度和强度都可能加大，而且不确定性和风险也将进一步增加。图为西北某处建有引水渠的农地无水灌溉，绿色即将消亡。

Irrational Utilization of Water Resources

Wasteful and disordered use of water resources are common and serious, such as water interception, flood irrigation, excessive pumping of groundwater and illegal discharge and recharge of sewage. These irrational uses of water resources give rise to many problems, including extinction of many wetlands, farmland abandonment, dead growth of trees, increasingly large underground funnels, abnormal microclimate and groundwater pollution etc. Currently, the water resources crisis has become the most serious crisis, and water challenge the biggest challenge. Global climate change has exacerbated China's water crisis and significant changes will happen to the pattern of water resources and water disasters. Not only the frequency and intensity of floods and droughts are likely to increase, the uncertainty and risk will also grow. The photo shows a farmland in northwest China, the green is dying out despite the diversion canal.

建设围垦　乱占湿地

城市建设、工程开发、农田围垦对湿地资源的过度开发和大量开垦，使湿地面积迅速减少，导致了湿地蓄洪防涝功能减低、鸟类栖息地遭受破坏。据不完全统计，新中国成立以来，长江流域有1/3以上的湖泊被围垦，因围垦而消失的湖泊达到1000多个，总面积达13000多平方千米，丧失湖泊调蓄洪水的容积大概与三峡大坝的蓄洪容积差不多，将近570亿立方米。滨海湿地更丧失了50%左右。图为渤海滩涂被一期接一期地围垦。

Reclamation of Wetlands

Excessive development and reclamation of wetland resources for such purposes as city construction, engineering development and farmland reclamation take a toll on wetlands area, undermining wetlands functions of flood storage and prevention, and the destruction of bird habitats. According to incomplete statistics, since 1949, more than one third of lakes in the Yangtze river basin have been reclaimed, and more than 1,000 lakes disappeared as a result of reclamation, with a total area of over 13,000 square kilometers, the flood retention volume lost thereby is almost the same as that of the three gorges dam, namely nearly 57 billion cubic meters. In addition, coastal wetlands have lost by around 50%. The photo shows Bohai mudflats being reclaimed time and again.

湿地污染日趋严重

湿地的主要污染物来自工农业生产排放的废水、居民生活污水以及农药、化肥、除草剂的不当使用等。许多天然湿地已成为工农业废水、生活污水的承泄区。这不仅使湿地水质恶化，而且对湿地的生物多样性造成严重危害，导致湿地质量下降，功能衰退。在中国东部，迅速发展的工业和农业占据了大量湿地，而它们排放的污水，污染了平原湖区80%以上的湖泊。图为小工厂排污对周围环境的影响。

Increasing Pollution in Wetlands

Major pollutants of wetlands come from industrial and agricultural wastewater, domestic sewage, and improper use of pesticides, fertilizers and herbicides. Many natural wetlands have become the drainage area for industry and agriculture waste water and sewage, which not only deteriorates the wetland water quality, but also seriously harms the wetland biodiversity, leading to the decrease of the wetland's quality and the decline of its functions. In east China, the rapid development of industry and agriculture has occupied a large area of wetlands, and sewage discharges of which polluted more than 80% lakes in the plain areas. The picture shows the influence of a small factory's pollution on the surrounding environment.

生物资源过度利用

在重要经济海区和湖河渔区滥捕乱捞现象相当严重，不仅使多种重要的天然鱼类资源受到损害，而且严重威胁着湿地中其他物种如鸟类、两栖爬行类的生存与发展。对湿地生物资源的过度开发利用，使湿地生物多样性衰退的趋势加剧，从而严重影响了湿地的生态平衡。据不完全统计，中国湿地濒危鸟类已经占亚洲濒危鸟类总量的一半还多，湿地高等植物也已经有大约 100 种高度濒危。图为从鄱阳湖区捞捕上来的鱼，可谓“断子绝孙”一网打尽。

Over Exploitation of Biological Resources

Serious over-fishing in important economic sea zone and lakes and rivers not only damages various important natural fish resources, but also poses a serious threat to the survival and development of other species in the wetland, such as birds, amphibians and reptiles. Biological resources in wetlands have been exploited and utilized excessively, and it exacerbating the declining trend of the wetland biodiversity, which staggeringly affected the wetland ecological balance. According to incomplete statistics, endangered wetland birds in China account for more than half of those in Asia, and around 100 species of wetland higher plants are highly endangered. The photo shows fish of different sizes being caught in one net in PoYang lake, without any consideration of breeding future generations.

外来物种入侵　生物多样性锐减

巴西龟、福寿螺、食人鲳、互花米草、凤眼莲等外来物种的侵入和泛滥，造成了对当地湿地生态环境的危害，不仅影响了原来的生物结构，而且水质加速恶化，生态系统遭到破坏，生物多样性锐减。在国际自然保护联盟公布的最具危害性的 100 种外来生物中，中国就有 50 多种，其中最严重的有 11 种，已经给我国造成每年大概 600 亿元的直接经济损失，生态方面的危害难以估量。图为滇池上大面积疯长的凤眼莲。

Sharp Species of Decrease Biodiversity due to Invasion of Alien Species

Alien species such as Brazilian tortoise, apple snails, Piranha, Spartina alterniflora Loisel, eichhornia crassipes reproduce in large quantity and invade local wetland ecological environment, which not only affects the original biological structures, but also accelerates deterioration of water quality, damaging the ecosystem and biodiversity. Among the 100 most dangerous alien invasive species disclosed by the World Conservation Union (IUCN), more than 50 species can be found in China, and 11 of which are most serious and have caused about RMB 60 billion of direct economic loss each year in China. The damage on ecological regard is beyond calculation. The photo shows excessive widespread of eichhornia crassipes in Dianchi lake.

成就和目标：

Achievements And Objectives:

湿地自然保护区建设

建立自然保护区是保护湿地自然生态系统最好、最有效的办法，中国从 1978 年建立了第一个湿地类型的自然保护区之后，迄今已建立了 577 处。其中最具代表性的是三江源国家级自然保护区，2000 年，在誉为“中华水塔”、“亚洲水塔”的青藏高原上，在长江、黄河、澜沧江源头地区建立了中国最大的湿地类型的自然保护区。目前，已有上百亿元的资金被用于三江源区的保护。图为时任中共中央总书记的江泽民为该保护区题写的碑名，这是我们党和国家最高领导人首次为自然保护区提名。

Wetlands Nature Reserve Establishment

Establishing nature reserves is the best and the most efficient way to protect the wetland ecosystem. China set up the first wetland nature reserve in 1978, and up until now, more than 577 wetland nature reserves have been established. Three-river Source National Nature Reserve, established in 2000, is the most representative one. Situated in the headwater region for three rivers, Yangtze River, Yellow River and Lantsang River, on the Qinghai-Tibet plateau known as "China's water tower" and "Asia's water tower", this nature reserve is China's largest wetland nature reserve. For the time being, investment for conservation purpose made in this reserve has reached more than 10 billion of Yuan. The picture shows then President Jiang Zemin's inscription for the reserve, this is the first time that the top leader of China inscribed for a nature reserve.

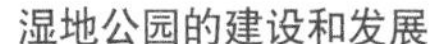

湿地公园的建设和发展

作为湿地保护体系的重要组成部分，在不适宜建立保护区的地方建立了湿地公园，以扩大湿地保护的面积。2005 年中国的第一个国家湿地公园在杭州西溪建立，迄今已经建有 429 处，总面积超过 180 万公顷。和保护区严格保护的方针不一样的是，湿地公园遵循的是"保护优先、科学恢复、合理利用、持续发展"的原则。既保护了湿地，又为民众增添了生态旅游、自然教育、休闲度假的场所。图为在中俄边界的黑瞎子岛上建立的湿地公园。

Establishment and Development of Wetland Parks

As an important part of wetland conservation system, wetland parks have been set up where the local conditions are unsuitable for establishing nature reserves, so as to enlarge the area of wetlands under conservation. In 2005 China's first national wetland park Xixi National Wetland Park was set up in Hangzhou. Until now, there are 429 wetland parks, with a total area of more than 1.8 million hectares. Unlike the rigid policies for wetland nature reserve, wetland parks follow the principles of "conservation first, scientific restoration, wise use and sustainable development" which conserves wetlands and at the same time provide spaces of nature education and recreation for the general public. The picture shows the wetland park established on Bolshoy Ussuriysky island on the Sino-Russian border.

国际重要湿地及国家、地方重要湿地的保护和建设

《湿地公约》的全称是《关于特别是作为水禽栖息地的国际重要湿地公约》，因此国际重要湿地的保护和建设不仅体现在中国这个大国履行国际义务的责任上，是全球生态系统的重要组成部分，也体现在对重要湿地保护管理的较高水平上。中国自 1992 年加入《湿地公约》后，已有 46 处湿地被列为国际重要湿地，在《中国湿地保护行动计划》中，有 173 处湿地列入国家重要湿地名录。目前，我国已初步建立了以湿地自然保护区为主体，湿地公园和自然保护小区并存，其他保护形式为补充的湿地保护体系。图为我国首批加入国际重要湿地的黑龙江扎龙国家级自然保护区。

Conservation and Development of Ramsar Sites and Wetlands of National and Local Importance

The full name for *Ramsar Convention* is *Convention on Wetlands of International Importance Especially as Waterfowl Habitat*, therefore conservation and development of Wetlands of International Importance (or Ramsar Sites) in China reflects not only that China, as a key part of global ecological system, is shouldering its international responsibility, but also its sound management on important wetland conservation. China joined the Ramsar Convention in 1992, ever since then, 46 wetlands have has been listed as Ramsar Sites, and 173 wetlands have been included in the list of wetlands of national importance according to *China Wetlands Conservation Action Plan*. For the time being, a wetlands conservation system with wetlands nature reserve as the main body, coupled with wetlands parks and small protected areas, and complemented by other forms of conservation has been initially established. The picture shows Heilongjiang Zhalong National Nature Reserve, one of the first Ramsar Sites of China.

湿地恢复

通过生态补水、湿地植被恢复、栖息地恢复、湿地污染控制、有害生物防治等工程措施，以逐步修复退化湿地的功能，改善湿地生态，维护湿地生态系统健康是湿地恢复的重要举措。如青藏高原高寒湿地、黑龙江三江平原湿地、长江中下游湖泊湿地等经过工程治理，湿地面积和功能得到了部分恢复。通过工程措施，广东湛江、海南东寨港、福建闽江河口、广西北仑河口等地的红树林也得到了很大程度的恢复。2006 年国家湿地恢复工程实施以来，全国每年新增湿地面积达 2 万多公顷。图为曾经干裂丧失现在正在恢复的湿地。

Wetlands Restoration

Through such engineering measures as ecological compensation, wetland vegetation restoration, habitat restoration, wetland pollution control and pest control, degraded wetland functions and wetland ecology could be gradually improved and the health of wetland ecosystem be maintained. For instance, through engineering management measures, the area and functions of alpine wetland on Qinghai-Tibet plateau, Heilongjiang Sanjiang plain wetland, and lake wetlands in the middle and lower reaches of the Yangtze River have been partially recovered Through engineering measures, mangrove forests in Zhanjiang of Guangdong, Dongzhaigang of Hainan, Minjiang Estuary of Fujian, and Beilun Estuary of Guangxi have been largely restored. Since the inception of national wetland restoration project in 2006, wetland area increases by more than 20000 hectares annually. The photo shows a previously dry and cracked wetland being restored.

可持续利用示范和能力建设

采用生态种植、生态养殖、退田（耕）还湿、退养还滩、退（禁）牧还湿等措施，逐步减轻湿地的开发利用程度，确保湿地资源的可持续利用。用示范点、示范基地等带动湿地周边的乡镇、市县。在工程建设中，通过开展持久的社区宣传教育和技能培训，使广大农牧民和渔民自觉地关心湿地、献策湿地、保护和合理利用湿地，进一步扩大了湿地保护的社会基础，为当地社会经济的可持续发展做贡献。图为科研技术人员在湿地做监测技能培训。

Demonstration and Capacity Building on Sustainable Use

Through such measures as establishing ecological plantations, developing ecological cultivation, converting cropland, aquaculture area and grazing land back to wetlands, and prohibiting grazing in wetlands, the intensity of wetlands utilization have been gradually reduced, in a bid to ensure the sustainable use of wetland resources. Demonstration sites have been established to showcase the villages, townships, cities and counties surrounding wetlands. By conducting community publicity and education and skills training on a regular basis, the awareness of the broad masses of farmers and fishermen on wetland conservation have been raised, who became consciously caring about wetlands and wisely using wetlands resources. Thus, the social basis for wetland conservation has been further expanded and more contribution has been made to the sustainable development of local economy. The photo shows a technician conducting skill training on monitoring in a wetland.

湿地保护建设远景目标

根据《全国湿地保护工程规划（2002～2030年）》等国家重大规划，到2030年，使全国湿地保护区达到713个，国际重要湿地达到80个，全国自然湿地保护率由目前的43.51%提高到2020年的70%，到2030年使90%以上天然湿地得到有效保护。完成湿地恢复工程140.4万公顷，在全国范围内建成53个国家湿地保护与合理利用示范区。建立起比较完善的湿地保护、管理与合理利用的法律、政策和监测科研体系。

要实行严格的湿地保护红线，到2020年中国湿地面积不得少于8亿亩；要建立湿地生态效益补偿机制；要实行严格的水资源总量控制及用水效率控制制度，万元国内生产总值和万元工业增加值用水量明显降低，农田灌溉水有效利用系数提高；要建立水功能区限制纳污红线，严格控制入河排污总量，到2020年全国主要江河湖库水功能区水质达标率要提高到80%左右；要把生态用水占总用水量的比重由目前的2%提高到2030年的8%。水土流失面积占国土面积的比重由2010年的37%，降至2030年的25%。

党的十八大明确要求扩大湿地面积，增强湿地生态系统稳定性，这是一项长期而艰巨的任务。经过持久不懈的努力，到2030年，我国江河湖库等湿地的水生态环境将得到显著改善，形成较为完整的湿地保护、管理、建设体系和河湖健康保障体系，重现湿地生命之水、生态之水、优美之水的景象，使我国进入湿地保护和管理先进国家的行列，为天蓝、水清、人幸福的“美丽中国”做出应有的贡献。

图为美丽清澈、水草肥美的巴音布鲁克湿地。

Long-term Objectives for Wetlands Conservation

According to such national plans as the *National Wetland Conservation Plan 2020-2030*, by 2030, the number of national wetland nature reserves will reach 713, wetlands of international importance 80, and the national natural wetlands conservation ratio will increase from the current 43.51% to 70% by 2020. By 2030, more than 90% of natural wetlands will be put under effective conservation. Some 1.404 million hectares of wetlands are planned to be restored, and 53 national demonstration sites for conservation and wise use will be established. A comparatively improved legal, policy and monitoring system will be established for wetland conservation, management and wise use.

The Red Line for wetlands conservation will be rigorously observed by, which means that by 2020, the area of wetlands should be no less than 800 million mu; a compensation system for ecological benefits will be set up. A stringent water management system for controlling total water resources and water use efficiency will be implemented which aims at significantly reducing water consumption per RMB 10,000 GDP and per RMB 10,000 industrial added value, and improving the water use efficiency of farmland irrigation by 2020. A red line for controlling the river pollution in water function area will be established to strictly limit the total amount of sewage discharged into rivers.By 2020, the rate for the water reaching quality standard in water function area of major rivers, lakes and reservoirs in China is to be increased to 80%. The proportion of water use for ecological purposes accounting for the total water consumption is to be increased from the current 2% to 8% by 2030. The proportion of water and soil erosion area to the total land area is to be reduced from 37% in 2010 to 25% by 2030.

The 18th National Congress of the Communist Party of China clearly required the widening of the wetlands area and to increase the stability of wetlands ecosystems. This is a long term arduous task. Through unremitting efforts, by 2030, wetlands ecological environment will witness a remarkable improvement, and a more complete system on wetland conservation, management and construction and a health security system for lakes and rivers will be formed. Due contributions will be made to build a more "beautiful China" with blue sky, clear water and joyful people by restoring the vitality, richness and beauty of wetlands and making China one of the advanced countries on wetland conservation and management.

The photo shows beautiful scenery on Bayanbulak Grassland wetland.

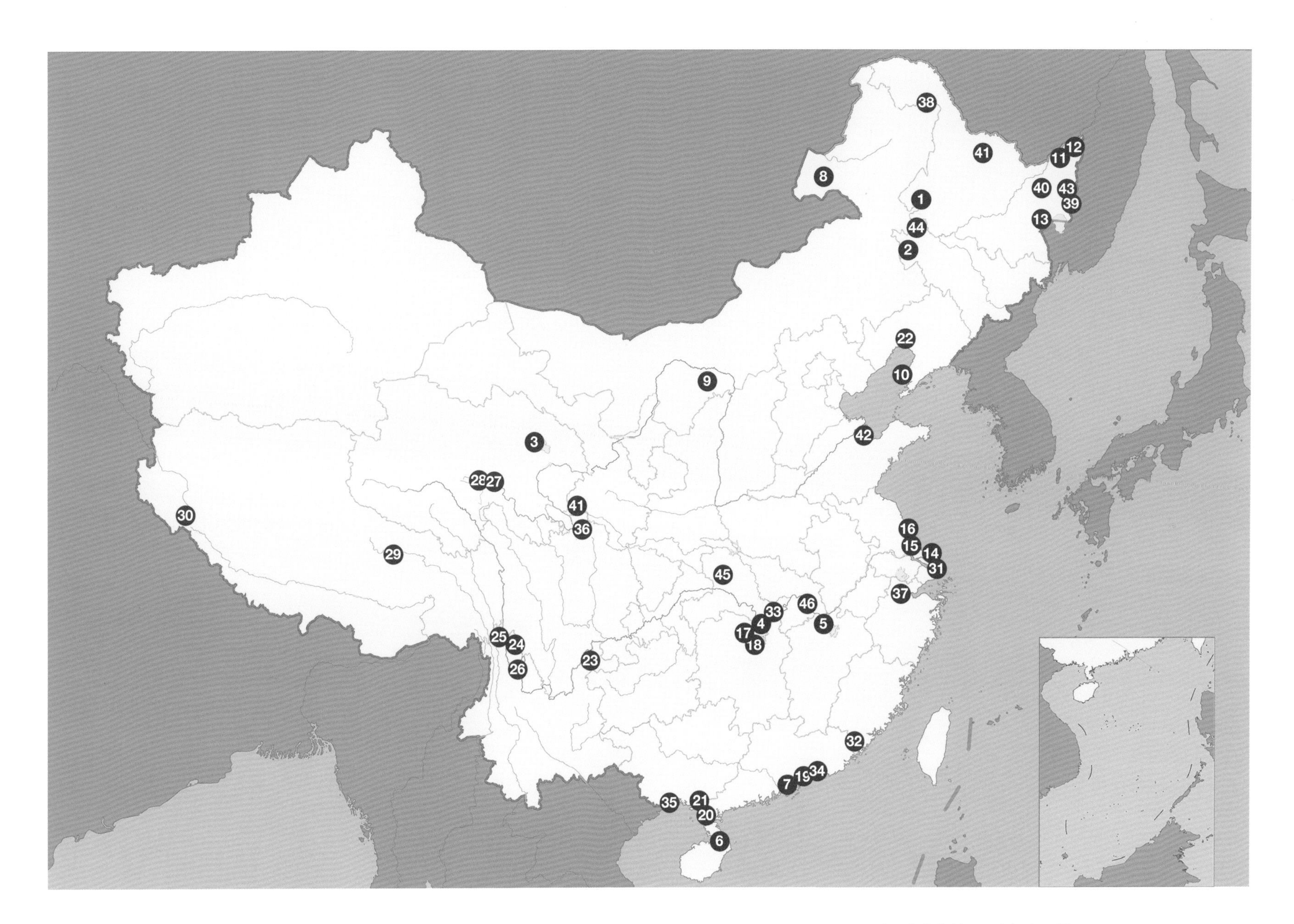

中国国际重要湿地名录

List of Wetlands of International Importance in China

括号中的页码表示该国际重要湿地的照片在本书中出现的页码。
The number in the parentheses indicates the Page Number where its photo appears.

❶ 黑龙江扎龙国家级自然保护区 (Page: 125, 132)
Zhalong National Nature Reserve of Heilongjiang Province

❷ 吉林向海国家级自然保护区 (Page: 134)
Xianghai National Nature Reserve of Jilin Province

❸ 青海青海湖鸟岛国家级自然保护区 (Page: 31, 52, 53)
Qinghai Lake Bird Islands National Nature Reserve of Qinghai Province

❹ 湖南东洞庭湖国家级自然保护区 (Page: 159)
Dong Dongting Lake National Nature Reserve of Hunan Province

❺ 江西鄱阳湖国家级自然保护区 (Page: 149, 156, 168)
Poyang Lake National Nature Reserve of Jiangxi Province

❻ 海南东寨港国家级自然保护区 (Page: 186)
Dongzhaigang National Nature Reserve of Hainan Province

❼ 香港米埔 – 后海湾湿地 (Page: 167)
Mai Po Marshes and Inner Deep Bay Wetland of Hongkong SAR

❽ 内蒙古达赉湖国家级自然保护区
Dalai Lake National Nature Reserve of Inner Mongolia Autonomous Region

❾ 内蒙古鄂尔多斯国家级自然保护区
Eerduosi National Nature Reserve of Inner Mongolia Autonomous Region

❿ 辽宁大连斑海豹国家级自然保护区
Da Lian Spotted Seal National Nature Reserve of Liaoning Province

⓫ 黑龙江三江国家级自然保护区 (Page: 121)
San Jiang National Nature Reserve of Heilongjiang Province

⑫ 黑龙江洪河国家级自然保护区
Honghe National Nature Reserve of Heilongjiang Province

⑬ 黑龙江兴凯湖国家级自然保护区
Xingkai Lake National Nature Reserve of Heilongjiang Province

⑭ 上海崇明东滩国家级自然保护区
Chongming Dongtan National Nature Reserve of Shanghai Municipality

⑮ 江苏盐城国家级自然保护区（Page: 160, 167）
Yancheng National Nature Reserve of Jiangsu Province

⑯ 江苏大丰麋鹿国家级自然保护区（Page: 167）
Dafeng Milu National Nature Reserve of Jiangsu Province

⑰ 湖南西洞庭湖自然保护区
Xi Dongting Lake Nature Reserve of Hunan Province

⑱ 湖南南洞庭湖湿地和水禽自然保护区
Nan Dongting Lake Wetland and Water Birds Nature Reserve of Hunan Province

⑲ 广东惠东港口海龟国家级自然保护区
Huidong Harbor Sea Turtle National Nature Reserve of Guangdong Province

⑳ 广东湛江红树林国家级自然保护区（Page: 187）
Zhanjiang Mangrove National Nature Reserve of Guangdong Province

㉑ 广西山口红树林自然保护区（Page: 177）
Shankou Mangrove Nature Reserve of Guangxi Zhuang Autonomous Region

㉒ 辽宁双台河口国家级自然保护区（Page: 130, 131）
Shuangtai Estuary National Nature Reserve of Liaoning Province

㉓ 云南大山包国家级自然保护区（Page: 111, 115）
Dashanbao National Nature Reserve of Yunnan Province

㉔ 云南碧塔海湿地
Bitahai Wetland of Yunnan Province

㉕ 云南纳帕海湿地
Napahai Wetland of Yunnan Province

㉖ 云南拉市海湿地
Lashihai Wetland of Yunnan Province

㉗ 青海鄂陵湖湿地（Page: 56）
Ngoring Lake Wetland of Qinghai Province

㉘ 青海扎陵湖湿地（Page: 36）
Gyaring Lake Wetland of Qinghai Province

㉙ 西藏麦地卡湿地
Maidika Wetland of Tibet Autonomous Region

㉚ 西藏玛旁雍错湿地
Manasarovar Wetland of Tibet Autonomous Region

㉛ 上海市长江口中华鲟自然保护区
Yangtze Estuary Chinese Sturgeon Nature Reserve of Shanghai Municipalty

㉜ 福建漳江口红树林国家级自然保护区（Page: 187）
Zhangjiang Estuary National Mangrove National Nature Reserve of Fujian Province

㉝ 湖北洪湖湿地（Page: 148）
Honghu Wetland of Hubei Province

㉞ 广东海丰湿地
Haifeng Wetland of Guangdong Province

㉟ 广西北仑河口国家级自然保护区（Page: 187）
Beilun Estuary National Nature Reserve of Guangxi Zhuang Autonomous Region

㊱ 四川若尔盖湿地国家级自然保护区（Page: 46, 50, 51）
Zoigê Wetland National Nature Reserve of Sichuan Province

㊲ 浙江杭州西溪湿地（Page: 170）
Hangzhou Xixi Wetland of Zhejiang Province

㊳ 黑龙江南瓮河国家级自然保护区（Page: 124）
Nanweng River National Nature Reserve of Heilongjiang Province

㊴ 黑龙江珍宝岛国家级自然保护区（Page: 128）
Zhenbao Island Wetland National Nature Reserve of Heilongjiang Province

㊵ 黑龙江七星河国家级自然保护区
Qixing River National Nature Reserve of Heilongjiang Province

㊶ 甘肃尕海湿地（Page: 83）
Gahai Wetland of Gansu Province

㊷ 山东黄河三角洲国家级自然保护区（Page: 139, 141）
The Yellow River Delta National Nature Reserve of Shandong Province

㊸ 黑龙江东方红湿地国家级自然保护区
Dongfanghong Wetland National Nature Reserve of Heilongjiang Province

㊹ 吉林莫莫格国家级自然保护区
Momoge National Nature Reserve of Jllin Province

㊺ 湖北神农架大九湖湿地
Shennongjia Dajiu Lake Wetland of Hubei Province

㊻ 湖北武汉沉湖湿地自然保护区（Page: 169）
Wuhan Chen Lake Wetland Nature Reserve of Hubei Province

指定国际重要湿地是履行《湿地公约》的重要内容，也是促进缔约国境内重要湿地保护管理的重大举措。按《湿地公约》规定，具有代表性、稀有性和独特性的自然或近自然湿地，或此湿地保护着易危、濒危、极度濒危物种种群或群落，或支持着动植物种生命周期的某一关键阶段等条件，就可由公约缔约方向公约秘书处提出申请，经核准后列入《国际重要湿地名录》。中国自参加《湿地公约》以来，已经有46块湿地列入《国际重要湿地名录》。

2000年11月8日，国务院17个部门共同参加并公布实施的《中国湿地保护行动计划》中，共列有173块国家重要湿地。按照国家林业局公布的、2013年5月1日起施行的《湿地保护管理规定》，国家重要湿地由国家林业局会同国务院有关部门制定标准和划定，并向社会公布。

这些国际重要湿地和国家重要湿地对于维持我国生物多样性，构建我国生态安全体系，实现社会和经济可持续发展起着重要的作用。

Designation of wetlands of international importance is an important obligation of contracting parties in implementing *Ramsar Convention*, and also a major approach of promoting wetland conservation and management within the territory of contracting parties. According to the Criteria for Identifying Wetlands of International Importance, if a wetland site contains a representative, rare, or unique example of a natural or near-natural wetland type found within the appropriate biogeographic region, or supports vulnerable, endangered, or critically endangered species or threatened ecological communities, or supports plant and/or animal species at a critical stage in their life cycles, or provides refuge during adverse conditions, the contracting parties then may submit an application to the Secretariat to be included into the *List of Wetlands of International Importance*. Since China entered the *Ramsar Convention*, a total of 46 wetlands have been designated as wetlands of international importance.

On November 8, 2000, 17 ministries under the State Council jointly issued the *China Wetlands Conservation Action Plan*, in which a total of 173 wetlands of national importance have been listed. Released by the State Forestry Administration on May 1, 2013, *Regulations on Wetland Conservation and Management* stipulates that the standards for identifying wetlands of national importance shall be formulated by the State Forestry Administration with relevant ministries under the State Council in cooperation, and the list shall be published to the public.

These wetlands of international importance and national importance are critical for maintaining biodiversity and building ecological security system in China, playing an important role in sustainable social and economic development.

物种简介

Introduction of Species

为了便于读者进一步了解物种情况和查阅，此部分列出了在本书中出现的物种，及其中文名、拉丁学名、保护级别、濒危程度等，还标上了该物种在书中出现的页码。

简写情况如下：

CHINA I = 国家一级保护野生动（植）物

CHINA II = 国家二级保护野生动（植）物

IUCN EW = IUCN物种红色名录等级，野外灭绝

IUCN CR = IUCN物种红色名录等级，极危

IUCN EN = IUCN物种红色名录等级，濒危

IUCN VU = IUCN物种红色名录等级，易危

IUCN NT = IUCN物种红色名录等级，近危

IUCN LC = IUCN物种红色名录等级，无危

CITES I = CITES附录 I 物种

CITES II = CITES附录 II 物种

CITES III = CITES附录 III 物种

注：

- 为保护野生动植物，维护生态平衡，我国先后颁布了《中华人民共和国野生动物保护法》和《中华人民共和国野生植物保护条例》，其中规定国家对珍贵、濒危的野生动植物实行重点保护，并根据物种的珍贵濒危程度和管理严格要求依次分为一级保护和二级保护（简写为CHINA I 和CHINA II）。

- 世界自然保护联盟（简称IUCN）是目前世界上最大的、最重要的世界性保护联盟。IUCN编制的濒危物种红色名录是被广泛接受和使用的受威胁物种分级标准体系。该组织每年评估数以千计物种的绝种风险，将物种编入9个不同的保护级别：依次为灭绝（EX）、野外灭绝（EW）、极危（CR）、濒危（EN）、易危（VU）、近危（NT）、无危（LC）、数据缺乏（DD）和未予评估（NE）。

- 《濒危野生动植物种国际贸易公约》，是全球缔约国之间为了保护野生动植物物种不至于由于国际贸易而遭到过度开发利用而进行的国际合作。《公约》将受管理的野生动植物物种按照其物种状况及其受贸易影响的严重程度依次列为附录I、附录II和附录III名单（简写为CITES I、CITES II和CITES III）。

This part has listed all the species introduced in the Album for reference. The introduction indicates its Chinese name, Latin name, grade of protection and degree of endangered condition. It also shows the number of pages where the specific species are introduced.

The abbreviations are as follows:

CHINA I = Wild Fauna and Flora under the 1st Grade National Protection

CHINA II = Wild Fauna and Flora under the 2nd Grade National Protection

IUCN EW = Extinct Species in the Wild of the IUCN Red List of Threatened Species,

IUCN CR = Critically Endangered Species of the IUCN Red List of Threatened Species,

IUCN EN = Endangered Species of the IUCN Red List of Threatened Species,

IUCN VU = Vulnerable Species of the IUCN Red List of Threatened Species,

IUCN NT = Near Threatened Species of the IUCN Red List of Threatened Species,

IUCN LC = Least Concern Species of the IUCN Red List of Threatened Species,

CITES I = Species listed in Appendices I of Convention on International Trade in Endangered Species of Wild Fauna and Flora (CITES)

CITES II = Species listed in Appendices II of Convention on International Trade in Endangered Species of Wild Fauna and Flora (CITES)

CITES III = Species listed in Appendices III of Convention on International Trade in Endangered Species of Wild Fauna and Flora (CITES)

Note:

In order to protect wild animals and plants and maintain the ecological balance, the Chinese government has successively promulgated the Law on the Protection of Wild Animals of the People's Republic of China and the Regulations on the Protection of Wild Plants of the People's Republic of China, which stipulate that the Chinese government protects the rare and endangered wild animals and plants, and categorized the species as CHINA I and CHINA II according to its degree of rareness and endangered condition, as well as the management requirement.

The World Conservation Union (IUCN) is currently the largest and most important world union for the protection of wild fauna and flora. The IUCN Red List of Threatened Species is a widely accepted and applied system. The IUCN carries out an annual evaluation on extinction risk of more than 1000 species and categorizes the species into nine different protection levels according to the evaluation: Extinct (EX), Extinct in the Wild (EW), Critically Endangered (CR), Endangered (EN), Vulnerable (VU), Near Threatened (NT), Least Concern (LC), Data Deficient (DD), and Not Evaluated (NE).

The Convention on International Trade in Endangered Species of Wild Fauna and Flora (CITES) is an international agreement between contracting parties aiming at ensuring that international trade in species of wild animals and plants does not threaten their survival. The species covered by CITES are listed in the appendices I , II and III (abbreviated as CITES I , CITES II and CITES III) according to the situation of the species and the degree of influence from trading.

白琵鹭
Platalea leucorodia
CHINA Ⅱ, IUCN LC, CITES Ⅱ
Page: 141

黑嘴鸥
Larus saundersi
IUCN VU
Page: 6

刺碱蓬
Suaeda glauca
Page: 6, 130

小天鹅
Cygnus columbianus
CHINA Ⅱ, IUCN LC
Page: 10, 139

普通鸬鹚
Phalacrocorax carbo
IUCN LC
Page: 31

黑颈鹤
Grus nigricollis
CHINA I, IUCN VU, CITES I
Page: 47, 50, 111, 115

赤麻鸭
Tadorna ferruginea
IUCN LC
Page: 47

藏野驴
Equus kiang
CHINA I, IUCN LC, CITES Ⅱ
Page: 49

棕头鸥
Larus brunnicephalus
IUCN LC
Page: 52

斑头雁
Anser indicus
IUCN LC
Page: 53

黑鹳
Ciconia nigra
CHINA I, IUCN LC, CITES Ⅱ
Page: 61

胡杨
Populus euphratica
Page: 68, 75

大天鹅
Cygnus cygnus
CHINA Ⅱ， IUCN LC
Page: 72, 82, 143

河狸
Castor fiber
CHINA I, IUCN LC
Page: 79

遗鸥
Larus relictus
CHINA I, IUCN VU, CITES I
Page: 80

蓑羽鹤
Anthropoides virgo
CHINA Ⅱ, IUCN VU, CITES Ⅱ
Page: 81

灰鹤
Grus grus
CHINA Ⅱ, IUCN LC, CITES Ⅱ
Page: 83

朱鹮
Nipponia nippon
CHINA I, IUCN EN, CITES I
Page: 89, 110

杉叶藻
Hippuris vulgaris
Page: 104

云南光唇鱼
Acrossocheilus yunnanensis
IUCN LC
Page: 105

钳嘴鹳
Anastomus oscitans
IUCN LC
Page: 108, 109

红嘴鸥
Larus ridibundus
IUCN LC
Page: 110, 114

大鲵
Andrias davidianus
CHINA Ⅱ, IUCN CR, CITES I
Page: 110

凤头麦鸡
Vanellus vanellus
IUCN LC
Page: 110

水雉
Hydrophasianus chirurgus
IUCN LC
Page: 110

小鸊鷉
Tachybaptus ruficollis
IUCN LC
Page: 110

牛背鹭
Bubulcus ibis
IUCN LC
Page: 121

丹顶鹤
Grus japonensis
CHINA I, IUCN EN, CITES I
Page: 132, 167

白枕鹤
Grus vipio
CHINA Ⅱ, IUCN VU, CITES I
Page: 133, 139, 156

泽鹬
Tringa stagnatilis
IUCN LC
Page: 137

鹤鹬
Tringa erythropus
IUCN LC
Page: 137, 158

反嘴鹬
Recurvirostra avosetta
IUCN LC
Page: 137

林鹬
Tringa glareola
IUCN LC
Page: 137

斑尾塍鹬
Limosa lapponica
IUCN LC
Page: 137

金眶鸻
Charadrius dubius
IUCN LC
Page: 137

黑翅长脚鹬
Himantopus mexicanus
IUCN LC
Page: 137

白腰杓鹬
Numenius arquata
IUCN NT
Page: 133

黑尾塍鹬
Limosa limosa
IUCN NT
Page: 137

狍子
Capreolus pygargus
IUCN LC
Page: 138

鸳鸯
Aix galericulata
CHINA Ⅱ
Page: 139

鸿雁
Anser cygnoides
IUCN VU
Page: 139

水獭
Lutra lutra
CHINA Ⅱ， IUCN NT, CITES I
Page: 139

须浮鸥
Chlidonias hybridus
Page: 139

疣鼻天鹅
Cygnus olor
CHINA Ⅱ， IUCN LC,
Page: 139

苍鹭
Ardea cinerea
IUCN LC
Page: 140

白鹤
Grus leucogeranus
CHINA I, IUCN CR, CITES I
Page: 156, 168

荇菜
Nymphoides peltatum
Page: 164

黄菖蒲
Iris pseudacorus
Page: 164

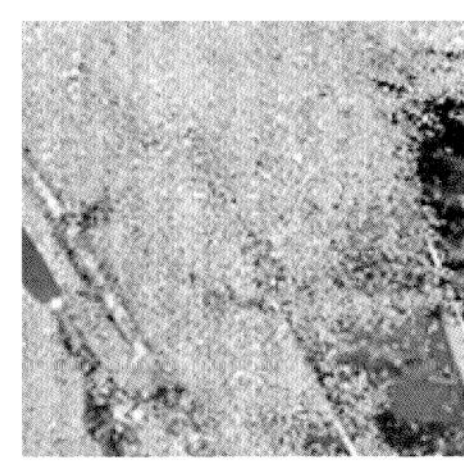

浮萍
Lemna minor
IUCN LC
Page: 164

菱角
Trapa bicornis
Page: 164

疏花水柏枝
Myricaria laxiflora
Page: 164

菹草
Potamogeton crispus
Page: 164

梭鱼草
Pontederia cordata
Page: 165

江豚
Neophocaena phocaenoides
CHINA Ⅱ， IUCN VU, CITES I
Page: 166

黑脸琵鹭
Platalea minor
CHINA Ⅱ， IUCN EN
Page: 167

麋鹿
Elaphurus davidianus
CHINA I, IUCN EW
Page: 167

白鹭
Egretta garzetta
IUCN LC
Page: 167

中华花龟
Mauiemys sinensis
IUCN EN
Page: 167

扬子鳄
Alligator sinensis
CHINA I, IUCN CR, CITES I
Page: 167

东方白鹳
Ciconia boyciana
CHINA I, IUCN EN, CITES I
Page: 169

草海桐
Scaevola sericea
Page: 177

招潮蟹
Uca pugnax
Page: 187

弹涂鱼
Periophthalmus cantonensis
Page: 187

海鸬鹚
Phalacrocorax pelagicus
CHINA Ⅱ， IUCN LC
Page: 189

翱翔飞鱼
Exocoetus volitans
Page: 190

红脚鲣鸟
Sula sula
CHINA Ⅱ, IUCN LC
Page: 191

玳瑁
Eretmochelys imbricata
CHINA Ⅱ, IUCN CR, CITES I
Page: 191

鲎
Limulus polyphemus
IUCN NT
Page: 191

银鸥
Larus argentatus
IUCN LC
Page: 192